Operational Design

Distilling Clarity from Complexity for Decisive Action

JEFFREY M. REILLY, PhD
Department of Joint Warfare Studies
Air Command and Staff College

Air University Press
Maxwell Air Force Base, Alabama

August 2012

Library of Congress Cataloging-in-Publication Data

Reilly, Jeffrey M.
   Operational design : distilling clarity from complexity for decisive action / Jeffrey
M. Reilly.
      p. cm.
   Includes bibliographical references and index.
   ISBN 978-1-58566-221-0 (alk. paper)
   1. Operational art (Military science) 2. Military planning. 3. Military art and
science—Decision making. I. Title.
   U162.R36 2012
   355.401—dc23

                                                                    2012025006

                        First Printing August 2012
                        Second Printing April 2017
                        Third Printing July 2018

Disclaimer

Air University Press
600 Chennault Circle, Building 1405
Maxwell AFB, AL 36112-6010
http://www.airuniversity.af.mil/AUPress/
Facebook: https://www.facebook.com/AirUnivPress
Twitter: https://twitter.com/aupress

Contents

Illustrations

Figures

Tables

About the
Author

Dr. Reilly is a retired Army officer with 26 years of active-duty service. He began his service as a draftee and served 28 consecutive months in Vietnam, Thailand, Laos, and Cambodia. His theater-level planning and operations experience includes serving as a theater-level combined and joint operations officer, Plans division chief, and member of the "two major theater war" plans team. He is an adjunct faculty member for the NATO School's Operational Planning Course, a frequent speaker at the USAF Weapons Instructor Course, and a member of the chairman of the Joint Chiefs of Staff's Military Education Coordination Council Working Group. Dr. Reilly has also given a number of presentations at international defense colleges, including the Ethiopian Defense Staff College in Addis Ababa and the Polish National Defense University in Warsaw. Additionally, he conducted research on design in Afghanistan during 2010 and 2011. He currently serves as director of joint education at the Air Command and Staff College.

Preface

Correct theories, founded upon right principles, sustained by actual events of wars, and added to accurate military history, will form a true school of instruction for generals. If these means do not produce great men, they will at least produce generals of sufficient skill to take rank next after the natural masters of the art of war.

—Antoine Henri de Jomini
The Art of War

The exploration of operational design is a transformative venture. It represents a tremendous leap in intellectual theory that will have far-reaching implications on the future of planning. In spite of its obvious importance, however, design remains shrouded in the inherent complexities that cloak the operational level of war and the concepts involved in campaign planning. Perhaps even more significant is the fact that we have collectively failed to empower the intrinsic relationship between design and decision analysis. The value of design is not just providing a mechanism to construct a campaign plan. Its true value rests in its fundamental capability to facilitate decision making. If approached deliberately, design provides a foundational basis for formulating preplanned decision points and creating the structure for linking decision analysis to emergent opportunities. Linking operational design with decision analysis substantially reduces the risk associated with an operation and increases the probability of a plan surviving first contact.

This analysis takes a didactic approach. It attempts to demystify the aura surrounding operational design by presenting a theoretical framework for comprehending its fundamental precepts. The goals of this analysis are threefold: provide a methodological example for understanding and applying design, show how design enhances decision making and risk analysis, and investigate the major differences between design in major combat operations and design in counterinsurgency (COIN).

The contents of this study should not be construed as either prescriptive or mechanistic. Warfare is a multifaceted entity conducted in evolving operational environments and against complex, adaptive adversaries. Design is not a sequential methodology or a simplistic checklist. It is a foundational part of operational art that provides the crucial element of structure. This heuristic examination of design simply searches for a way to explain design's intricate structural relationships and highlight the intrinsic potential for deliberately crafting decision analysis. When commanders and staffs approach operational design from this perspective, we move one step closer to the natural masters of the art of war.

Illustrations were created by the author unless otherwise noted.

Chapter 1

The Process of Operational Design

Operational design's principal purpose is to distill clarity from complexity for decisive action. This factor is frequently overlooked in the intellectual discourse that surrounds the exploration of design. The reasons for this include a tendency to focus on the complexity of operational environments and a reticence to link the process of design to a structure focused on decision making. Operations invariably occur in multifaceted environments. No commander ever has perfect information, all of the resources he or she desires, or enough time. The process of operational design, however, is not about discovering complexity. Design is about creating operational vision from complexity and offsetting the uncertainty embedded in operations with effective decisions.

Although there are no precise formulas for developing design,[1] attaining effectiveness requires a structural basis that joint force commanders (JFC) and their staffs can understand and apply. Many contemporary theorists, however, refuse to accept structure as a key component of design. They feel structure leads to prescription and diminishes the intellectual exploration of each operation's unique mission requirements and environmental characteristics. This perception has a degree of validity, but it is also myopic. As long as JFCs and their staffs recognize the iterative nature of planning, differences in mission requirements, and distinctive environmental anomalies, structure is an invaluable organizational tool.

Structure simply provides the basic parameters for focusing critical thinking and guiding design's intellectual discourse toward key planning goals.[2] When adequate structure is absent, intellectual discussions founder and critical thinking disintegrates. This impairs the ability to judiciously assess the environment, identify the problem requiring resolution, and develop an operational approach that supports commander-level decision making.

Decision making is centrifugal to design's structure because it determines who has the initiative and ultimately who wins. Operational environments are constantly in a state of change, and when staffs assist commanders in understanding decision criteria and risk mitigation, the commanders are much more likely to act decisively. However, when the design process lacks the structure to focus on decisions, commanders can fall prey to excessive analysis and potentially catastrophic forms of indecision. A classic illustration of this is Gen George B. McClellan's Peninsula Campaign during the American Civil War.

At 2:00 a.m. on 22 July 1861, less than a day after the Union's humiliating defeat at Manassas, Virginia, a telegram directed General McClellan to take command of the Army of the Potomac.[3] McClellan was far from an average officer. He had studied under Prof. Dennis Hart Mahan at the US Military Academy, was versed in Jomini, and in 1846 graduated second in a class of 59 cadets. He also participated in Gen Winfield Scott's amphibious landings at Vera Cruz during the Mexican War, and in 1855 he was one of three American officers sent to Europe to study the Crimean War. McClellan left the Army in 1857 and by 1860 was the president of the Ohio and Mississippi Railroad. When the Civil War began, he was credited with crushing organized southern resistance in northwest Virginia through the engagements at Rich Mountain and Corick's Ford. In terms of organizational and administrative skills, combat experience, and intellect, McClellan personified the exact type of leader needed to regalvanize and lead a dispirited force.

Between July and October 1861, McClellan meticulously organized and trained a fighting force of 120,000 men. McClellan, however, was reluctant to use this force and act decisively. He believed that the Confederate forces confronting him outside Washington were numerically superior to the army he had forged. To assess the Confederate strength, McClellan authorized Allan Pinkerton to conduct reconnaissance missions and espionage behind Confederate lines. Pinkerton estimated that the Confederate forces consisted of 115,000 men and over 300 pieces of artillery. In fact, the Confederate forces led by Gen Joseph E. Johnston numbered fewer than 45,000 at their peak—less than 40 percent of what McClellan believed was opposing him. By January 1862 President Lincoln was totally dismayed with McClellan's lack of initiative, exclaiming, "If General McClellan does not want to use the Army for some days, I would like to borrow it provided I could see how it could be made to do something."[4] President Lincoln and many of the other Union leaders pressed McClellan to mount an overland campaign to seize the Confederate capital of Richmond. McClellan resisted, however, because in his estimation an overland campaign would entail attacking over 100 miles through hostile terrain, crossing multiple rivers, and maintaining a tenuous supply line. To McClellan's credit he developed one of the American Civil War's most imaginative and ambitious campaign plans. The plan focused on leveraging the Union's naval superiority and achieving an asymmetric advantage over the Confederate forces. Originally known as the Urbana Plan, it called for bypassing the Confederate forces entrenched around Manassas and Centerville and conducting an amphibious assault to seize Richmond. President Lincoln had several major reservations about the risk involved in the plan, and he subjected it to several military councils. However, when McClellan persisted in advocating his plan, President Lincoln eventually gave it tacit approval.

On 7 March 1862, however, Gen Joseph E. Johnston, forewarned of McClellan's plan, repositioned the Confederate forces around Manassas to positions south of the Rappahannock River. When the Confederates vacated their defensive positions, reporters went forward and discovered Quaker guns (wooden logs set up to replicate cannons) and defenses that were far from impregnable. Undeterred, McClellan revised his plan and sought approval to land Union forces farther south in the vicinity of Fort Monroe. On 17 March 1862, a Union naval force of 400 ships began transporting 121,500 men, 44 artillery batteries, and over 15,000 horses to the Virginia peninsula (fig. 1).[5]

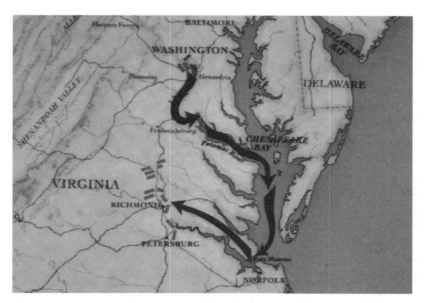

Figure 1. Peninsula Campaign, 1862. (Courtesy of Florentine Films)

By early April, McClellan had amassed approximately 55,000 soldiers in front of the Confederate defenses at Yorktown. But again McClellan was reticent to act. His plan had been disrupted by not receiving command of the 12,000-man force at Fort Monroe and the loss of McDowell's Corps, which was redirected to protect Washington, DC.

In the meantime, the Confederate forces under Gen John B. Magruder employed a series of deception operations that convinced McClellan the Confederates had a much larger force. In actuality, Magruder had a force of fewer than 13,000 men stretched along a 12-mile line, and his superior General Johnston considered the defenses untenable. On 3 May 1862, before McClellan could attack, Johnston began withdrawing the Confederate forces and ordered James Longstreet to conduct a delaying action near Williamsburg. On 15 May at Drewry's Bluff, Confederate forces defeated an attempt by McClellan to conduct a simultaneous attack using Union naval forces along the James River. This tactical setback created a major psychological disruption for McClellan.

As McClellan's lead forces came within six miles of Richmond, the Chickahominy River overflowed its banks and split the Union forces before they could attack. The Confederates seized this opportunity and attacked McClellan at Seven Pines on 31 May. The battle resulted in 790 Union soldiers killed, 3,594 wounded, and approximately 647 captured. Confederate losses included 980 killed, 4,749 wounded, and 405 missing.[6] The most significant casualty was Gen Joseph E. Johnston. When General Johnston was wounded, Robert E. Lee took command of the Army of Northern Virginia. Although McClellan subsequently had numerous opportunities to seize the initiative, he persisted in waiting for reinforcements and optimal weather conditions.

McClellan's campaign culminated in the Seven Days' Battles from 25 June to 1 July 1862. The Union and Confederate forces fought six major engagements during this time, and the only Union defeat was at Gaine's Mill. At the battle of Malvern Hill on 1 July 1862, the final battle of the campaign, Lee's forces conducted a series of disjointed frontal assaults on Union defensive positions and lost heavily. Confederate losses exceeded 5,300 casualties. In spite of this, McClellan refused to counterattack and instead withdrew to Harrison's Landing on the James River. The campaign losses for both sides were appalling. The Confederates suffered 20,614 casualties, and the Union forces 15,849. On 3 August 1862 McClellan's forces began withdrawing from the peninsula, thus ending a campaign that could have substantially altered Southern resistance. The effects of McClellan's campaign, however, were not isolated to the Virginia peninsula. When General Lee understood that McClellan's forces were withdrawing, he committed his forces to destroy the Union forces under General Pope at the Second Battle of Manassas on 29 and 30 August 1862.

In retrospect, McClellan's Peninsula Campaign was a viable plan. McClellan created a sophisticated plan that used an indirect approach, exploited the Union's naval superiority, and focused on a key decisive point, the Confederate capital of Richmond. McClellan demonstrated exceptional innovation in logistics and naval gunfire support and even attempted to use two observer balloons to provide intelligence on enemy movements. His primary faults, however, revolved around over-analyzing his adversary's capabilities, failing to take decisive action, and not conducting a realistic risk assessment. Collectively these omissions made McClellan extremely vulnerable to not one but a series of Confederate deception operations. Although Alan Pinkerton's inflated intelligence summaries are partially to blame for this, the Confederates had exhibited a clear pattern of using deception to mask their limited military capacity. They used deception throughout 1861 and 1862 in their defenses immediately outside Washington, DC, at Manassas, at Centerville, and again during the Peninsula Campaign at Yorktown. The crucial liability in McClellan's campaign design was in McClellan's own inflexibility and indecisiveness. The outcome of his plan might have been dramatically different if he had incorporated better decision making and risk analysis tools in his design.

Understanding the Operational Environment

The process of operational design is an analytical fusion of strategic direction, the operational environment, and the problem requiring resolution. The result of design is the development of an operational approach that engenders flexibility through incisive decision making and balanced risk analysis. Understanding the environment is the initial step and foundational basis for the entire design process. The environment is a multifarious, interactive, and constantly evolving series of systems. It encompasses not only the immediate area of operations, but also all areas, actors, and factors that either influence or have the potential to influence the area of operations. The concept of understanding the environment is not new, but the methodology has been modified to support a more holistic critical analysis of the environment's systemic interconnectivity. Understanding the environment also yields a structural origin for envisioning key aspects of design such as end-state conditions, objectives, centers of gravity (COG), decisive points, lines of operation, and adversary perspectives.

The construct of framing the environment consists of two interrelated subsystems that foster a shared understanding of the environment's interconnectivity and a mechanism for identifying the problem requiring resolution (fig. 2). The two subsystems are the observed system and the desired system.[7] The observed system is an analytical depiction of the environment as it currently exists. It consists of regularly interacting, interdependent, and independent elements that affect the joint operations area and ultimately the mission. Planners begin framing the environment by examining key factors such as principal actors and their interrelationships; cultural relationships; historical context; physical geography; instruments of power; elements of power; and political, military, economic, social, information, and infrastructure (PMESII) elements.

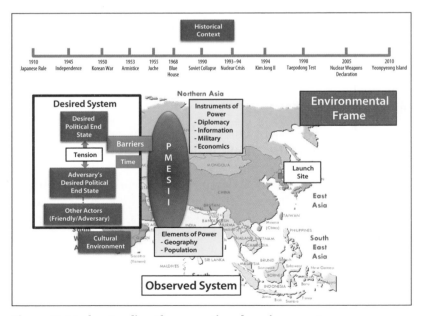

Figure 2. Understanding the operational environment

The Observed System

There is no exact alchemic prescription for visualizing the observed system because the observed system is a construct of what one has the capability to see and understand. The term "observed system" does not necessarily convey that what one is observing is the system as it actually exists. Not all variables are readily visible or have evolved sufficiently to be systemically linked. Some operational environments evolve over time and may not be self-evident. An illustration of this is Operation Iraqi Freedom during the period 2003–2006. Iraq was engulfed in almost uncontrollable sectarian violence, complex interactions among tribes, clans, and other groups, and international interference by Iran. Not until late 2005 and early 2006 could crucial patterns of activity be identified to create the foundation for the allied surge's success.

Another critical realization is that not all observed systems are openly accessible and access to an environment's systemic information is not always equal. As a result, the accuracy of a depiction of an observed system is dependent upon two essential variables: access to information and the ability to critically analyze that information. An example of this is the Democratic People's Republic of Korea (DPRK). Most nations have very limited access to North Korea, and this directly affects what can be observed and mapped within the DPRK's environmental context. Concepts such as the DPRK's *juche* are superficially understood by the West but have a tremendous impact on North Korea's national "will to act." First proclaimed by Kim Il-sung in 1955, *juche* has evolved as part socio-political and part religious philosophy that justifies major policy decisions.[8] *Juche* serves as an unknown determinant in what the West perceives as irrational acts of provocation, such as the sinking of the South Korean ship *Cheonan* in March 2010 and the shelling of Yeonpyeong Island in January 2011.

The ability to critically analyze information within the observed system is equally as important as access to information. When planners develop the observed system, they must be able to analyze available information in depth and comprehend what the information implies for operations. For example, the current illiteracy rate in Afghanistan is approximately 72 percent. Although this seems like an innocuous bit of information for an underdeveloped nation, it has a major effect on operations and mission success. Illiteracy affects operations in Afghanistan in several important ways. First, illiteracy is the single biggest obstacle undermining the building of effective Afghan security forces. Only 14 percent of the individuals entering the Afghan armed forces are literate.[9] Security forces cannot be effective unless they can read and write orders and understand basic manuals associated with tactics, administration, equipment, and maintenance. Literacy is an important enabler to professionalize security forces, reduce corruption, enhance stewardship, and increase recruiting.[10] The magnitude of Afghan illiteracy presents an incredible challenge for the NATO Training Mission–Afghanistan, and it dramatically slows the ability to field Afghan National Army and Afghan National Police forces. The NATO Training Mission has made literacy a priority, but as of May 2011, only 50 percent of the armed forces were projected to be literate at the first-grade level by January 2012.[11]

The second way illiteracy affects operations in Afghanistan is in information operations (IO). If the majority of the population cannot read or write, the mechanism used to communicate IO has to be modified to accommodate that factor. Traditional IO methodologies do not work among illiterate populations. In October 2010 the International Council on Security and Development (ICOS) conducted a survey of 1,500 men in northern and southern Afghanistan. Sixty-eight percent of the respondents in the Helmand and Kandahar provinces had seen photos of the planes striking the Twin Towers on 9/11. However, 92 percent of those respondents could not relate the photos to the events of 9/11 or identify the role those events played in international intervention.[12] Additionally, it is important to stress that our adversaries understand how to leverage illiteracy to support their IO. The traditional agent of change in Afghanistan is the mullah. However, many mullahs are either illiterate or functionally illiterate. The Taliban exploit this by providing interpretations of written documents such as the Koran to justify acts of violence.

However, perhaps the most significant implication illiteracy has for operations is on the future of a nation. In today's international environment, no nation can compete effectively or be self-sufficient without an educated workforce. Illiteracy also implies that decades of investment will be required to create that workforce, which should be accounted for in the development of end-state conditions.

The intent of highlighting illiteracy is not to dramatize the effects of illiteracy but to underline the need to critically analyze all variables identified in the observed environment. Access to information is important. However, understanding what the information means is a vital element in projecting the accuracy of what planners observe. In many instances military planners will simply not have the skill required to interpret key variables. Accurately depicting the observed system requires an expansive diversity of expertise. It is virtually impossible for military planners to be experts in areas such as international economics, foreign relations, forensics, and ethnography. Consequently, it is essential to integrate other government agencies, nongovernmental organizations, and specialists as early as possible into the design process. It is crucial to meticulously describe the observed system as precisely as possible if planners are going to leverage a genuine understanding of the desired system's potential.

The Desired System

The environment's desired system analyzes the perceived tension between the strategic political direction provided by national or multinational authorities and the adversary's desired political end state. This analysis examines not only the tensions between friendly and adversary political end-state conditions but also time, other actors capable of influencing the end state, and barriers to the operation.

A common misperception in the dissection of friendly and adversary desired political end states involves the intricate nature of what an actor wants to do and what an actor is willing to do. A subtle but significant difference between these two contrasts affects the derivation of key planning assumptions, branches, and sequels. An illustration of this is China's theory of unrestricted warfare. In February 1999 senior Chinese colonels Qiao Liang and Wang Xiangsui published *Unrestricted Warfare* in response to perceptions about US global power projection. *Unrestricted Warfare* advocates going beyond traditional boundaries when necessary to achieve national political objectives. Qiao and Wang describe unrestricted warfare as the use of "supra combinations" that transcend the confines of the military sphere:

> These things make it clear that warfare is no longer an activity confined only to the military sphere, and that the course of any war could be changed, or its outcome decided, by political factors, economic factors, diplomatic factors, cultural factors, technological factors, or other nonmilitary factors. Faced with the far-reaching influence of military and non-military conflicts in every corner of the world, only if we break through the various kinds of boundaries in the models of our line of thought, take the various domains which are so

completely affected by warfare and turn them into playing cards deftly shuffled in our skilled hands, and thus use beyond-limits strategy and tactics to combine all the resources of war, can there be the possibility that we will be confident of victory.[13]

This text signals a willingness to use cyber warfare, information operations, and terrorism to attack both military and nonmilitary targets. These targets would include financial institutions, power grids, water supplies, and other key infrastructure components. In Qiao's words, "The first rule of unrestricted warfare is that there are no rules, with nothing forbidden."[14]

As planners deliberate aspects of what an adversary wants to do and is willing to do, they must be cognizant of time. The temporal dimension has an effect far beyond being a limited resource. Planning in a time-compressed environment is extraordinarily difficult, but time is also crucial in determining how long it will take to achieve the desired political end state and how long the effects must last. When national or multinational authorities promulgate strategic direction, they must have an acute awareness of the "will" to achieve the end state and the conditions necessary to maintain that "will." Military operations are expensive in terms of lives and financial burden, and there is a direct correlation between the level of ambition and cost. As the cost of a military operation goes up, the level of ambition and "will" goes down. This is a common phenomenon associated with most wars, and it is evident in the support for operations in Afghanistan. In 2011 the Congressional Research Service estimated the cost of Operation Enduring Freedom at approximately $2 billion per week.[15] This financial burden, combined with almost 3,000 International Security Assistance Force (ISAF) members killed in action, has caused a marked decrease in US and coalition support for the war. A CNN survey conducted in the United States between 17 and 19 December 2010 revealed that 63 percent of those surveyed opposed the war. This is not meant to judge operations in Afghanistan but simply to emphasize the fact that planners must understand the impact of time on national and coalition support for military operations.

Another aspect of time planners must consider is the relationship between the desired political end state and how long the conditions established by military operations will last. A phrase often used to describe political end state conditions is "long-term regional stability." The problem is that *long-term* is never adequately defined. Does *long-term* mean 10 years, 20 years, or longer? The answer has a major impact on the effectiveness of military operations and the development of termination criteria. On 27 July 1953 an armistice terminated major hostilities on the Korean peninsula. The armistice created a 160-mile-long and 2.5-mile-wide demilitarized zone (DMZ) between the Republic of Korea and the DPRK. When the armistice was signed, North Korea was not able to bring significant effects on the Republic of Korea. However, by the early 1990s advances in artillery systems, surface-to-surface missiles, and the DPRK's nuclear program drastically altered regional stability. It is crucial to grasp the fact that long-term regional stability does not necessarily constitute an indefinite state of being.

Just as time affects the desired system, other actors have a dynamic capability to influence the political end state. As planners develop the construct for the desired system, they should assess the possible effects—both positive and negative—of all actors on desired end-state conditions. Some actors will be allies, some overt adversaries, some neutral, and some neutral with the potential to intervene. It is absolutely imperative to map actor relationships, understand their systemic links, and develop actions that will set the conditions for achieving the political end state. Mapping actor relationships assists planners with identifying strategic and operational assumptions and guides the development of associated branches and sequels. Additionally, it provides the strategic-level vision for whole-of-government, intercombatant command, and multinational coordination.

Any investigation of systemic actor relationships (fig. 3) must also include an examination of how the actors make decisions. Decision-making theory is supported by many studies that include the rational actor, cognitive, cybernetic, and polyheuristic models. However, actors draw from a diverse set of frames of reference to make decisions, and there is no universal decision-making pattern. The key is recognizing the differences in the patterns and their impact on operations. Of course, an ally's decision making can have just as dramatic an effect on an operation as an adversary's.

In March 2002, the planning for Operation Iraqi Freedom hinged on the creation of a northern front that would attack Iraq through Turkey. By early 2003, however, US equipment was sitting on ships circling in the eastern Mediterranean Sea, waiting for the outcome of negotiations with the Turkish government.[16] On 1 March 2003 the Turkish Parliament decided in a 264–251 vote to veto the deployment of an estimated 62,000 US personnel onto Turkish soil to attack Iraq.[17] This

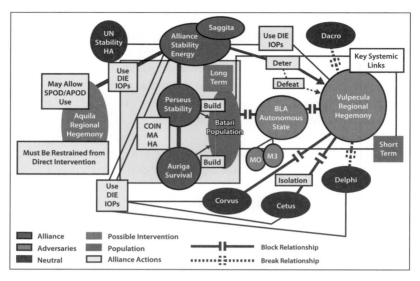

Figure 3. Mapping actor relationships and systemic links

demand probably did not seem unreasonable to most US military personnel because the United States was offering $30 billion in grants and loan guarantees and putting pressure on Europe to accept Turkey into the European Union. But without intimate knowledge of Turkey, its domestic politics, and its strategic concerns, this proposal was extraordinarily unrealistic. The outcome was not only a denial of using Turkey as a northern front, but also severely strained US-Turkish relations.

The last segment of analyzing the desired system is identifying barriers to the desired political end state. Determining barriers facilitates analysis of a critical aspect of planning that is often missing: expectation management. When a plan is formulated or undergoes a major revision, political leaders must understand the full scope of the plan's capabilities and its limitations. This fosters a more realistic examination of planning assumptions and promotes better decision making and risk analysis. Barriers exist in many forms, including time, force levels, interagency disputes, intergovernmental organizations, and numerous other key factors. As planners identify barriers, they must conduct a thorough assessment of the barriers' impacts on the desired political end-state conditions and raise critical issues to the political authorities for decisions. A realistic appraisal of barriers assists JFCs and their staff with conveying an accurate depiction of what can and cannot be accomplished, and it frames expectation management for political leaders. It also provides JFCs with a basis to make effective recommendations when political leaders direct a change in operational-level resources.

In August 2009, GEN Stanley McChrystal dramatically revised Operation Enduring Freedom's strategy. He transformed the operational focus to population-centric operations, revitalized theater command and control, created an ISAF operational headquarters, and pushed for an accelerated growth of Afghan security forces. The new strategy, however, was confronted by a number of operational limitations. The environment was almost totally devoid of critical logistics infrastructure, and the adversary was increasingly emboldened, aggressive, and adaptive. As a result, General McChrystal requested 40,000 additional personnel to execute the strategy. But when President Obama reviewed General McChrystal's strategy, he cut the requested forces by 10,000 personnel. This change altered the focus of the strategy; however, it maintained essentially the same objectives and political expectations. Political leaders expected the revised strategy to have the same striking success as the 2007 surge in Iraq. By July 2010 political leaders were requesting metrics to assess the strategy's effectiveness, in spite of the fact that the strategy's forces had not even fully closed in the theater. The result was a political miscue that emboldened insurgent efforts. The president announced his intent to initiate a withdrawal of forces in 2011 without meeting the operational conditions necessary for a withdrawal. This message was later corrected, but to the Taliban, Haqqani network, and other insurgent groups, it was an informational windfall. The point is that barriers and expectations must be reassessed during any reframing of the environment or revision of the strategy and not just during the development of the strategy.

The process of examining barriers during the analysis of the desired system is directly linked to correctly identifying the problem requiring resolution. As obvious as it seems, operational limitations significantly affect the ability to solve the problem identified by political leaders. Excessive operational limitations may not only prevent mission success but also precipitate a new set of problems that political leaders are unprepared to deal with or accept. Consequently, correctly identifying the core problem that actually requires resolution is a centrifugal component of the design process.

Identifying the Problem Requiring Resolution

One of the most important parts of the analysis of the observed and desired systems is the identification of the specific problem requiring resolution. Identifying the problem and understanding its surrounding context are vital to determining how to solve the problem. This step assists planners with recognizing what is possible and what is not possible. Additionally, it mitigates the possibility of solving the wrong problem.

Identifying the problem requires a comprehensive exploration of the contextual and environmental factors causing the tension between the desired political end states of friendly parties and those of adversaries. Isolating the underlying causes of tension empowers a much more precise methodology of what needs to be acted upon to achieve the desired political end state. To do this effectively planners must understand the potential to change the environmental conditions, the operational limitations, and the full implications of changing the environmental system. Grasping the environment's potential is important because the action taken may not only solve the problem but also prevent future problems. Identifying operational limitations is crucial to determining such factors as prohibited and required military actions, rules of engagement, diplomatic support agreements, and host nation requirements. Changing any contextual feature of the environment has second- and third-order effects on the environmental system. Planners need to project the potential implications of desired changes and screen those changes for possible threats to US national, coalition national, and regional national interests.

When analysis of the problem is completed, the result should be a clear, concise problem statement identifying the problem being solved and how it will be resolved. Of course the entire design process is iterative, and the problem statement may require several revisions to reach maturity. An example of an initial problem statement might be:

> Country X's overt military actions in the Caspian region threaten access to vital energy resources and international economic stability. The coalition seeks freedom of access to key energy resources and the establishment of the conditions necessary for long-term regional stability. Country X must be deterred from further aggression or risk coalition military intervention. Coalition forces will defend the territorial integrity of Country Y and ensure international access to energy resources.

Frame-of-Reference Theory

Before JFCs and their staff begin the development of an operational approach, they must validate their frame of reference for both the operational environment and problem identified. Frames of reference are double-edged swords. All individuals naturally use schemata to frame contextual elements and problems (fig. 4).[18] Consequently, in complex environments there is an innate tendency to focus on systemic similarities of previously learned schemata rather than identifying and investigating systemic anomalies. Isolating and examining the operational environment's systemic anomalies is an intrinsic element of forming schemata. The challenge is to ensure that the schemata used in the design process are correct.

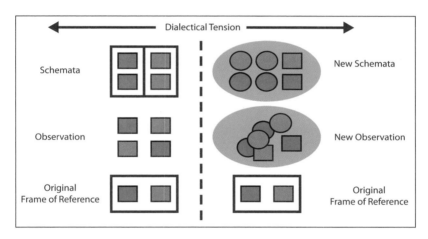

Figure 4. Frame-of-reference theory

One of the most common mistakes planners make is "mirror imaging," especially in unfamiliar cultural contexts. When the mind confronts new surroundings, it imposes schemata on the incoming sensory data. Noted developmental psychologist Jean Piaget theorized that the mind either assimilates new sensory data into existing schemata or modifies the schemata to incorporate the new data.[19] Both of these constructs have a certain basis of validity, but identifying the best way to combine data and schemata is situation-dependent. The key is to understand the dialectical tension between these two constructs and how they empower critical thinking skills.

An illustration of mirror imaging is an American attempting to comprehend Iran's national-level decision making (fig. 5). To most Americans, Mahmoud Ahmadinejad is the president of Iran and as such is Iran's principal decision maker. The American democratic ideal influences this perspective because we view officials elected by the electorate as the key components of the national decision-making system. As a result, we might perceive Iran's president, Parliament, and Assembly of Experts as the integral elements of all systemic political decision making. The danger of this point of view is it relegates key Iranian appointed officials such as the supreme leader, Guardian Council, Expediency Council, and Cabinet to mere influencers.

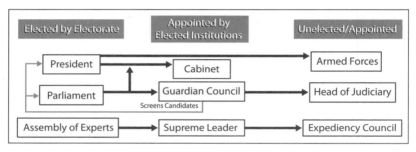

Figure 5. Perception of Iranian decision making

Iran's political decision-making system possesses a far more intricate nature. Figure 6 provides a cursory depiction of just how complicated Iranian decision making is in reality. Iran is a theocratic democracy, and the supreme leader, Ayatollah Ali Khamenei, exerts a central role in all decision making. As the supreme leader, Ayatollah Khamenei (not the president) is commander-in-chief of the armed forces, and he alone has the power to declare war, neutrality, or peace. Additionally, the supreme leader appoints and can dismiss the supreme commander of the Islamic Revolutionary Guard Corps, the leaders of the judiciary, and even the state media networks. He also has the authority to appoint six of the Council of Guardians' 12 members. This council is a dominant force in the oversight of Iranian parliamentary activities and determines which candidates are qualified to run for public office. The supreme leader's power base is amplified in two other significant ways. He controls the Expediency Council, which mediates disputes between Parliament and the Council of Guardians, and he has approximately 2,000 clerical field operatives positioned throughout the government.[20]

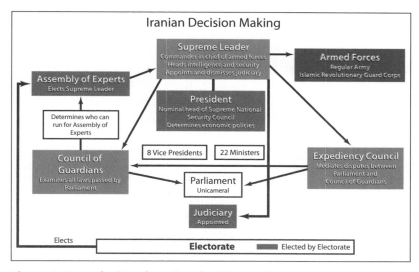

Figure 6. Complexity of Iranian decision making

When properly understood, these details provide a much different perspective on Iranian planning, decision, and execution cycles. When planners use frames of reference, they must ensure the frames are accurate visualizations or at least acknowledge key information voids and adjust their schemata. Thus it is important to leverage interagency analysis and seek assistance from specialists.

Understanding Gestalt Theory

Another critical aspect of understanding frames of reference is Gestalt theory. Christian von Ehrenfels and other German psychologists such as Max Wertheimer used Gestalt theory to explain perceptual patterns of wholeness and totality. This concept attempts to describe how people organize visual elements as a whole and stresses that what they actually see may not be the sum of the parts. In essence, Gestalt theory questions the derivation of patterns of wholeness and advocates a deeper critical analysis of perceptions. In Max Wertheimer's words, "Here too we find science intent upon a systematic collection of data, yet often excluding through that very activity precisely that which is most vivid and real in the living phenomena it studies. Somehow the thing that matters has eluded us."[21]

The significance of Gestalt theory is that as we develop the framework for the operational environment's observed and desired systems, we must be cognizant of different ways to visualize systemic characteristics. Figures 7 and 8 show examples of this. In figure 7, one individual may see a vase, and another may see two heads facing each other. In figure 8, a person with a macro perspective sees the letters *L* and *H*. However, a person with a micro perspective may see a collection of small *H*s forming the letter *L* and small *L*s forming the letter *H*. One of these patterns represents reality, and the other is an aberration.

Figure 7. Different perceptual patterns

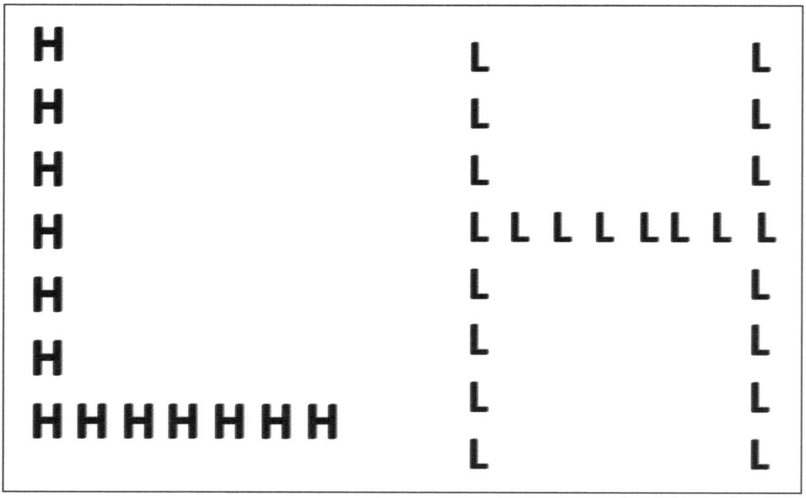

H	L	L
H	L	L
H	L	L
H	L L L L L L L L	
H	L	L
H	L	L
H H H H H H H	L	L
	L	L

Figure 8. Macro and micro perceptual patterns

Gestalt analysis emphasizes the criticality of viewing the environment from multiple perspectives to form a metacognitive understanding of the environment's complexity and interconnectivity. JFCs and their staffs must understand this because it has a tremendous effect on the development of the operational approach. If the perceived observed and desired systems are inaccurate, the discrepancies will have a cascading impact on the operational approach's end-state conditions, objectives, centers of gravity, decisive points, lines of operation, phasing, and assumptions.

Notes

(All notes appear in shortened form. For full details, see the appropriate entry in the bibliography.)

1. de Czege, "Systemic Operational Design," 2.
2. This monograph defines critical thinking as a judicious evaluation of commonly accepted beliefs and assumptions.
3. McPherson, *Battle Cry of Freedom*, 348.
4. Stackpole, *From Cedar Mountain to Antietam*, 1.
5. Sears, *George B. McClellan*, 167–69.
6. Johnson and Buel, eds., *Battles and Leaders of the Civil War*, vol. 2, *The Struggle Intensifies*, 219.
7. Joint Warfighting Center Doctrine Pamphlet 10, *Design in Military Operations*, 8.
8. Ch'oe, Lee, and de Bary, eds., *Sources of Korean Tradition II*, 419.
9. US Government Accountability Office, *Afghanistan Security*, 24.
10. US Department of Defense, *Report on Progress toward Security and Stability in Afghanistan*, April 2011, 20.
11. Mora, "Half of Afghan Military Forces Won't Achieve 1st Grade Literacy Level by 2012."
12. International Council on Security and Development, *Afghanistan Transition*, 26–28.
13. Qiao and Wang, *Unrestricted Warfare*, 191.
14. Ibid., 2.
15. See Belasco, *The Cost of Iraq, Afghanistan, and Other Global War on Terror Operations since 9/11*, 28–31.
16. Kapsis, "The Failure of U.S.-Turkish Pre-War Negotiations."
17. See Mintz, "How Do Leaders Make Decisions?," 8.
18. See Bartlett, *Remembering*.
19. See Piaget and Inhelder, *Memory and Intelligence*.
20. Maleki, "Decision Making in Iran's Foreign Policy." See also Afrasiabi and Maleki, "Iran's Foreign Policy after 11 September."
21. Wertheimer, "Gestalt Theory."

Chapter 2

Developing an Operational Approach

The operational approach is the "commander's description of the broad actions the force must take to achieve the desired military end state."[1] When JFCs and their staffs initiate the development of an operational approach, they must recognize the fundamental differences between the functions of operational design and the functions of operational art. The function of operational design is important because it determines how design relates to operational art and integrates into the joint operations planning process. Design is part of operational art, but the elements of design operate as discrete entities during the development of an operational approach. Effective planning and problem-framing relies on an unbiased approach to mission analysis. When operational art dominates the elements of design too early in the planning process, it unduly influences the process with predisposed courses of action (COA). This stifles critical thinking and prevents the development of innovative solutions to problems.

The operational approach to design is not new. It is as old as warfare itself. When man first used a stick to sketch an operational concept in the dirt, he developed the ability to cognitively link objectives with maneuver. In 1479 BC one of Egypt's greatest pharaohs, Thutmose III, used operational design to subdue a rebellion of Canaanite kings and princes at Megiddo.[2] Generally regarded as the first recorded battle in history, Megiddo provides crucial insights into the development of operational design and the integration of operational art. Based on what we know about Thutmose's campaign, his operational design reflected a remarkably sophisticated analysis of end-state conditions, objectives, effects, center of gravity, assumptions, and phases.[3]

The battle at Megiddo originated in a Canaanite quest to topple Egyptian regional domination. Encouraged by Egyptian domestic difficulties, the king of Kadesh formed a coalition of states to contest Egypt's hegemony. The coalition selected the key terrain in and around the walled fortress of Megiddo to defend its interests. This terrain included the vital lines of communication running from Egypt to Palestine, Syria, and Mesopotamia. When the Egyptians discovered this threat to their national interests, their army conducted a phased march of approximately 230 miles to the Yehem crossroads. At Yehem, Thutmose III held a war council to assess intelligence gathered on the Canaanite dispositions and to determine the best course of action. Thutmose had three possible axes of advance to attack the Canaanite coalition (fig. 9). The first was to the north toward Yokneam. This axis was defended by coalition infantry. The second was in the center leading directly into Megiddo. It was lightly defended with nearby chariots, but the terrain was restrictive and not favorable for deploying forces if attacked during the march. The third was to the south, and like the north axis, it was defended by infantry.

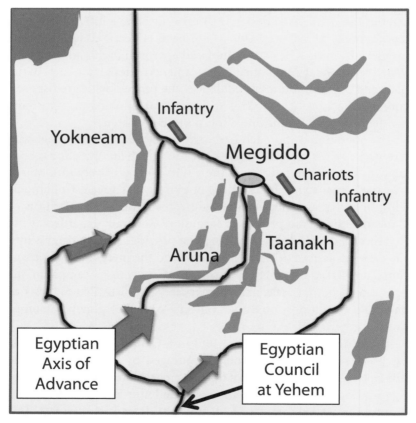

Figure 9. Egyptian courses of action at the Battle of Megiddo in 1479 BC. (Based on Yohanan Aharoni, *Carta's Atlas of the Bible* [Jerusalem: Carta, 1964]: 32.)

During the mission analysis, Thutmose's generals were predisposed to eliminate any COA associated with the center axis of advance. They begged Thutmose not to select the center route because of the restrictive terrain. Egyptian records reflect this opposition: "Let our victorious lord proceed according to the design of his heart therein, but do not cause us to march upon this impassable road."[4]

Thutmose, however, overruled his generals' objections. Based on intelligence reports, he had an accurate assessment of where the Canaanites had deployed their forces. He determined that attacking along the center axis would surprise his adversaries and allow him to crush the decisive point of Megiddo. Leading the Egyptians himself, Thutmose attacked along the center axis, catching the Canaanite coalition by surprise. This allowed Thutmose to deploy his forces in front of Megiddo before the Canaanite forces could redeploy and engage the Egyptians. When the two forces finally met, the Egyptians routed the Canaanites so fast that their retreating forces had to be hauled up the walls of Megiddo because the gates to the city were locked shut. The Egyptian attack, however, bogged down when the Egyptian forces stopped to pillage the coalition's encampment. This critical delay afforded the Canaanite forces the opportunity to escape into the fortress of Megiddo, robbing the Egyptians of a quick, decisive victory. The Egyptians laid siege to the city for five months, and eventually Megiddo surrendered. In spite of the lengthy siege, Thutmose achieved his desired strategic end-state conditions. His reign was never challenged by another rebellion, and the defeat of the coalition's forces forced other towns in the Syrian zone of influence to submit to Egyptian hegemony.[5]

Since Megiddo, the basic intellectual framework that military planners have used throughout history has changed very little. The contemporary explanation of operational design, however, remains unclear. Current joint doctrine establishes a clear purpose for operational design, but the analytical process underpinning joint planning remains vague.[6]

There are two principal reasons for this. First, although operational design is primarily an intellectual exercise based on experience and judgment, it has a defined structural component. Second, joint doctrine does not distinguish operational design and operational art as two distinct entities. In 1995 the keystone document Joint Publication (JP) 3-0, *Doctrine for Joint Operations*, introduced the 14 facets of operational art. In subsequent iterations of joint doctrine, these facets evolved into the 13 elements of operational design.[7] As a result, operational design and art have become blended with little explanation of their inherent distinctions. The current JP 3-0, *Joint Operations*, and JP 5-0, *Joint Operation Planning*, define operational design and operational art as follows:

> [Operational design is] **the conception and construction of the intellectual framework that underpins joint OPLANs and their subsequent execution.** . . .
>
> *Operational design* extends operational art's vision with a creative process that helps commanders and planners answer the ends–ways–means–risk questions.[8] (emphasis in original)
>
> [Operational art is] the cognitive approach by commanders and staffs—supported by their skill, knowledge, experience, creativity, and judgment—to develop strategies, campaigns, and operations and organize and employ military forces by integrating ends, ways, and means.[9]
>
> The thought process helps commanders . . . visualize how best to effectively employ military capabilities to accomplish their mission.[10]

These definitions present operational design and art as two separate but related functions. However, current doctrine does not delineate which of these elements relate to design and which relate to art. Using the above definitions, operational design elements should assist commanders and their staffs in creating a planning framework, and operational art elements should support the development of strategy during the COA-development step of the joint operation planning process (JOPP). Figure 10, based on JP 5-0, comes very close to illustrating this separation.

Elements of Operational Design	
Termination	Direct and indirect approach
Military end state	Anticipation
Objectives	Operational reach
Effects	Culmination
Center of gravity	Arranging operations
Decisive points	Forces and functions
Lines of operation and lines of effort	

Figure 10. Elements of operational design. (Based on JP 5-0, *Joint Operation Planning*, 11 August 2011, III-18.)

Why is it so important to separate these two concepts? The principal reason is to have a focused organizational structure that unambiguously articulates the preliminary vision for the operational approach. Dividing operational design and art into two separate entities provides commanders and their staffs with an impartial organizational structure for problem framing during the initiation and mission-analysis steps of the JOPP. Prematurely introducing complex operational art concepts such as anticipation, forces and functions, and operational reach into the mission-analysis step detracts from problem framing by developing a solution before the full nature of the problem has been determined. The significance of this organizational shift to separate operational design and art is that it reduces the tendency to interject preconceived solutions and strategies into the JOPP's mission-analysis step. This approach maximizes the potential for developing unbiased COAs. Once operational design creates the framework for informed vision, planners can use strategy to integrate operational art into the JOPP's COA-development, analysis, and comparison steps.

JP 1-02, *Department of Defense Dictionary of Military and Associated Terms*, defines strategy as a prudent idea or set of ideas for employing the instruments of national power in a synchronized and integrated fashion to achieve theater, national, and/or multinational objectives. Strategy provides a crucial mechanism for integrating operational design and art by outlining the planning guidance for COA development. Figure 11 shows a methodology for integrating operational design and art into the JOPP. This methodology uses operational design to begin structuring the operational approach in an unbiased environment during the JOPP's initiation and mission-analysis steps. Once the initial operational approach is framed, the JFC can introduce strategy to integrate operational art into the structure established by the elements of operational design. The challenge presented in our current joint doctrine is determining which elements of operational design support problem framing and which support strategy development.

An analysis of the 13 elements of operational design and the requirements for developing an operational approach reveal eight interrelated elements that provide a basic cognitive framework for problem framing. These elements are end state, objectives, effects, centers of gravity, decisive points, lines of operation, arrangement of operations, and assumptions. Seven of these eight elements are listed

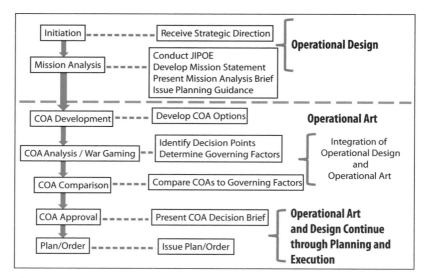

Figure 11. Separation of operational design and art in the JOPP

in joint publications as doctrinal elements of operational design. The only element not so listed is assumptions. Assumptions are included here as an element of operational design since they provide the core framework for all plans and are arguably the primary reason for most plans' failures. Assumptions help frame what a plan is going to do and what it is not going to do. They articulate major decision points and catastrophic risk and identify the branches and sequels required to mitigate risk.

In contrast to the design elements, the operational art elements focus on the development of strategy. They integrate with operational design during the JOPP's COA-development step to communicate the organization and employment of forces. These elements assist JFCs in formulating an initial strategy and providing guidance to the staff for COA development during the JOPP.

This deliberate bifurcation of operational design and art simplifies their integration into the JOPP by delineating key responsibilities along the lines of problem framing and military strategy development (fig. 12). This action does not relegate operational art to a subordinate position below operational design. In actuality, operational design is subordinate to operational art (fig. 13). When national or multinational leaders provide strategic direction, JFCs and planners

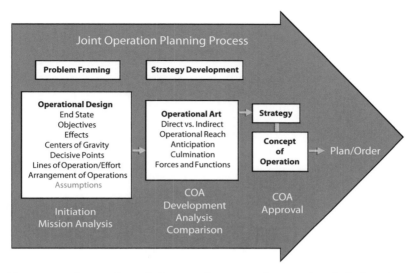

Figure 12. Separation of the 13 elements of operational design into distinct operational design and art elements and their relationship to the JOPP

use operational art to build the design structure. This structure then provides the analytical basis for understanding the environment, identifying the problem, and developing an operational approach. Design assists JFCs with creating a cognitive map to visualize the theater strategy and provide the guidance necessary for COA development. JFCs and planners use operational art to develop COAs based on the vision embodied in the commander's design. As planners analyze and war-game COAs, they can identify key preplanned decisions that leaders must make. These activities also present a focal point to formulate emergent opportunities that exploit projected environmental conditions or enemy actions. This deliberate integration of operational art and design forms a critical foundation for examining risk, comparing COAs, and selecting a COA. When a COA is finally selected, the JFC uses operational art and design to refine the cognitive map and solidify the operational vision that synchronizes tactical actions to accomplish operational objectives.

Recently several eloquent articles have described the complex origins and interactions among strategy, operational art, and design. However, these intellectual endeavors have little utility unless com-

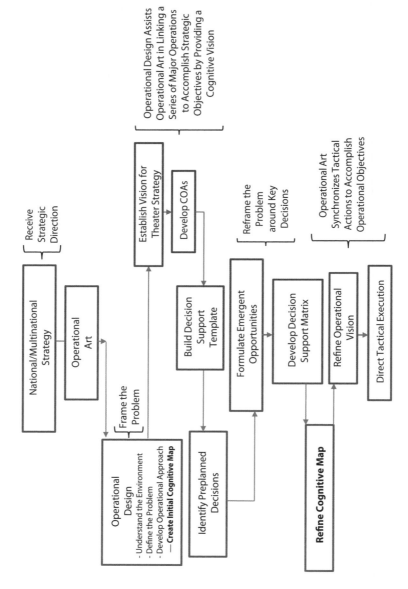

Figure 13. Relationship between operational art and design

manders and their staffs have a design methodology they can both understand and use. As this analysis continues, it describes *a* methodology that can be used to comprehend design and employ it to provide clarity in visualizing the operational approach.

Notes

1. Joint Publication (JP) 5-0, *Joint Operation Planning*, 11 August 2011, III-5.

2. For a description of Megiddo, see Fuller, *A Military History of the Western World*, vol. 1, 4–7.

3. Gabriel and Boose, *Great Battles of Antiquity*, 53.

4. Nelson, "Battle of Megiddo," 22. See also Pritchard, *Ancient Near Eastern Texts Relating to the Old Testament*, 234.

5. Gabriel and Boose, *Great Battles of Antiquity*, 59.

6. Dickson, "Operational Design," 23.

7. JP 3-0, *Doctrine for Joint Operations*, 1 February 1995, III-10.

8. JP 3-0, *Joint Operations*, 11 August 2011, II-4.

9. JP 5-0, *Joint Operation Planning*, 11 August 2011, GL-13.

10. Ibid., I-5.

Chapter 3

Methodology for Understanding and Employing Operational Design

A number of methodologies of varying complexity and focus explain how to construct an operational design for a campaign plan. Examples include the effects-based approach to operations, systemic operational design, and numerous fusion models. The methodology described in this analysis rests on the eight interrelated design elements discussed in chapter 2. It is presented in a sequential format based on conventional operations to provide foundational knowledge for understanding operational design's basic framework. *However, it should be emphasized that this methodology is not prescriptive. Operational design is an iterative process and not necessarily a sequential methodology.* Additionally, different types of operations have distinctive effects on how JFCs and their staffs develop supporting operational designs to frame problems and enhance decision making.

One of operational design's key products is a cognitive map. This map assists JFCs and their staffs with envisioning an entire operation. It also allows them to assess the congruency of critical links among the national strategic end state, military objectives, effects, centers of gravity, decisive points, and lines of operation.

The cognitive map evolves through several iterations as the JOPP progresses. During the JOPP's initiation and mission-analysis steps, the JFC develops a basic map that outlines the initial campaign strategy and provides guidance to the staff on course-of-action development. As the vision for a campaign unfolds during the JOPP's course-of-action analysis, comparison, and selection steps, a more detailed map should emerge that depicts phases and even stages within phases. There is no specified format for the cognitive map. The format should support the commander's ability to visualize the entirety of the campaign or operation. It should also assist the JFC in identifying critical decision points and serve as a tool to communicate his vision to civilian leaders, allies, and component commanders. Additionally, the format should aid the staff in checking the links among the elements of operational design and ensure the best possible alignment and arrangement of operations for accomplishing the national strategic end state. Figure 14 gives an example of a simplified cognitive map and the alignment of key elements of design.

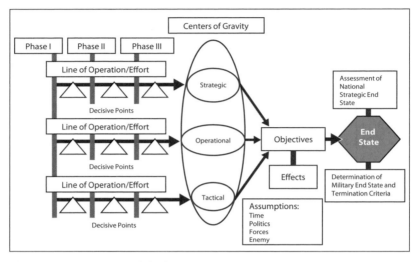

Figure 14. Operational design's cognitive map

Because there is no standardized interagency planning process, the cognitive map also provides a valuable communication mechanism for interagency coordination. Arraying the elements of operational design in a cognitive framework conveys key military operational concepts in a format that diverse agencies can grasp relatively quickly. It also presents a methodology for integrating the four national instruments of power linking the strategic and operational levels of operations, as well highlighting interagency decision points (fig. 15).

Understanding End State, Objectives, Effects, and Tasks

The point of origin for developing an operational approach is an analysis of strategic guidance and a comprehension of the national strategic end state. Joint Publication (JP) 1-02 defines *end state* as "the set of required conditions that defines achievement of the commander's objectives."[1] The gravity of employing the military instrument of power, however, mandates a clear understanding of strategic purpose bounded by a national strategic end state and a military end state. The president and the secretary of defense provide strategic guidance es-

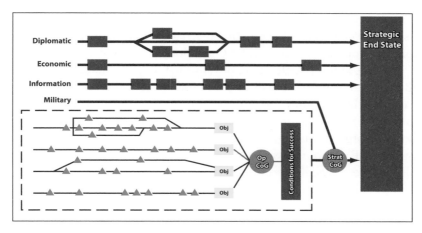

Figure 15. The cognitive map and the four national instruments of power

tablishing a set of national strategic objectives that should culminate in the accomplishment of the national strategic end state. The national strategic end state describes the president's political, informational, economic, and military vision for the region or theater when operations conclude.[2] These broadly expressed conditions provide the basis for determining the military end state and the termination criteria necessary for concluding military activities.

However, the derivation of strategic guidance is extraordinarily complex and dynamic. Guidance originates from numerous sources, the most common of which are strategic-level policy statements. These statements include the National Security Strategy, National Defense Strategy, National Military Strategy, Guidance for the Employment of the Force, Joint Strategic Capabilities Plan, and national security presidential directives. For example, the 2008 National Defense Strategy outlines five objectives:

- Defend the Homeland
- Win the Long War
- Promote Security
- Deter Conflict
- Win our Nation's Wars[3]

Strategic guidance can also be derived indirectly from interagency and even international directives, such as United Nations (UN) Security Council resolutions. Additionally, strategic direction can evolve out of executive-level addresses to the nation. One example of the latter is the development of the national strategic objectives forming the original end-state conditions for Operation Enduring Freedom (OEF).

On Tuesday, 11 September 2001, the history of our nation changed. Between the hours of 8:45 and 10:00 a.m., terrorists hijacked four commercial airplanes and crashed two into the World Trade Center, one into the Pentagon, and one into a field in rural Pennsylvania. Pres. George W. Bush articulated the end-state conditions for the ensuing OEF in a joint session of Congress on 20 September. Those conditions were broadly described as the destruction of terrorist training camps and infrastructure within Afghanistan, the capture of al-Qaeda leaders, and the cessation of terrorist activities in Afghanistan.[4] By the time the first strikes of OEF occurred on 7 October, the end-state conditions had expanded. On that date, Secretary of Defense Donald Rumsfeld amplified the president's guidance at a Department of Defense news conference.[5] He stated that the operation's objectives were to convince Taliban leaders that the harboring of terrorists is unacceptable, acquire intelligence on al-Qaeda and Taliban resources, develop relations with groups opposed to the Taliban, prevent the use of Afghanistan as a safe haven for terrorists, and destroy the Taliban military, allowing opposition forces to succeed in their struggle. Also, military force would help facilitate the delivering of humanitarian supplies to the Afghan people. The initial phase of OEF demonstrates the acute situational awareness JFCs and their staffs must have to plan effectively. Planners must understand that end-state conditions are dynamic and often emanate from nontraditional sources.

Another illustration of the complexity involved in assessing and understanding the national strategic end state is the development of objectives for Operations Desert Shield and Desert Storm. Initially, when Pres. George H. W. Bush addressed Congress on 5 August 1990, three days after Iraq's invasion of Kuwait, the US national policy objectives in the Persian Gulf were to:

- Bring about the immediate, complete, and unconditional withdrawal of all Iraqi forces from Kuwait
- Restore Kuwait's legitimate government
- Ensure the security and stability of Saudi Arabia and other Persian Gulf nations
- Ensure the safety of American citizens abroad

By 15 January 1991, on the eve of initiating air operations, these objectives had changed dramatically because of Iraq's refusal to withdraw from Kuwait and the UN's approval to use force. The national strategic end-state conditions evolved into:

- Destroy Iraqi capability to produce and employ weapons of mass destruction

- Destroy Iraqi offensive military capability

- Cause the withdrawal of Iraqi forces from Kuwait

- Restore the legitimate government of Kuwait[6]

During the iterative planning process, military planners have an obligation to seek clarification on unclear or ambiguous national strategic objectives that form the basis for the desired end state. Planners must assess the plausibility of attaining strategic, operational, and tactical objectives and understand exactly what they mean. For instance, if an objective says "reduce weapons of mass destruction," does this mean reduce weapons of mass destruction by a percentage, or destroy their employment and production capabilities? Another example is "halt enemy forces." Does this mean simply stop the enemy forces, or is the expectation to bring the enemy to a culmination point where he can no longer mount offensive operations or attain his objectives? The repercussions of misunderstanding the desired strategic end state are dramatically significant because of its link to the military end state and other corresponding objectives.

The military end state is military specific. It describes the national strategic end-state conditions that the military instrument of power will direct its efforts to achieve. The military end state is a set of required conditions that define the achievement of all military objectives. It normally represents a point at which the president no longer requires the military instrument of national power as the primary means to achieve remaining national objectives.[7] The military end state reflects the conditions established by the national strategic end state; however, it is more specific and normally incorporates supporting conditions, such as a UN Security Council resolution, associated with the other national instruments of power.

Although the termination of military operations is a political decision, effective planning cannot occur without a precise vision of the military end state and its associated termination criteria. Termination criteria are "the specified standards approved by the President and/or the Secretary of Defense that must be met before a joint operation can be concluded."[8] Knowing when to terminate military operations and how to preserve achieved advantages is essential to achieving the national strategic end state. The conditions, derived from the military end state and national strategic end state, contribute to developing termination criteria that must be met before concluding military operations. Termination criteria require planners to consider a wide variety of operational issues and anticipate the nature of post-conflict operations.[9] Termination criteria are dynamic. They evolve throughout the JOPP and are usually not evident until after the JOPP's course of action analysis and war-gaming step. Therefore, it is probably more efficient to use the military end-state conditions as the initial termination criteria until the COA analysis is complete.

Planners should also realize that termination criteria can change significantly during the execution of an operation because national or multinational leaders may opt to modify directed strategic end-state conditions to pursue less aggressive goals or take advantage of environmental opportunities. As a result, termination criteria are essentially in draft form until the strategic leadership directs the cessation of military hostilities. Once cessation is directed, planners need to review the termination criteria and refine them to meet leader expectations and environmental conditions. Table 1 shows an example of the relationships among a national strategic end state, a corresponding military end state, and termination criteria.

The analysis of the national strategic end state, military end state, and termination criteria, in conjunction with an assessment of the operational environment, provides the basis for determining objectives. Objectives are "clearly defined, decisive, and attainable goal[s] toward which every operation is directed."[10] They describe the political, economic, informational, and military goals necessary to accomplish the desired end state. Examples of military objectives are:

- Deter Vulpecula's forces from armed intervention into the country of Auriga
- Enable humanitarian assistance in Auriga
- If deterrence fails, defend Auriga from Vulpeculan intervention

JFCs and their staffs support and clarify objectives by developing measurable, desired strategic and operational effects and assessment indicators. An effect is a physical and/or behavioral state of a system

Table 1. Relationships among national strategic end state, military end state, and termination criteria

National Strategic End State	Military End State	Termination Criteria
International peace and security in the Zoran Sea region restored with full implementation of all UN Security Council resolutions (UNSCR) and the sovereignty and territorial integrity of Auriga preserved	• Peace and security of Auriga restored • UNSCR 1783 implemented, restoring a secure environment for the provision of humanitarian assistance • Regional peace and security no longer threatened by armed conflict • WMD threat eliminated	• Preconflict sovereignty and borders of Auriga intact • The flow of humanitarian aid into and evacuation out of Auriga uninterrupted by armed attacks • Vulpecula no longer able to mount offensive operations • Vulpecula's weapons of mass destruction (WMD) employment, production, and storage capabilities eliminated

that results from an action, a set of actions, or another effect. Effects are conditions derived from objectives and exist in two forms, desired and undesired. Desired effects describe the conditions necessary within the operational environment for achieving an objective. Undesired effects reflect the conditions to be avoided because they would inhibit the accomplishment of an objective. This analysis increases operational- and tactical-level understanding of the operation's purpose and the commander's intent. Additionally, this alignment provides the basis for operational-level assessment. If the effects correctly describe the achievement of the objective, JFCs have a mechanism for assessing the effectiveness of their operations. The conditions signified in the effects aid in determining tasks for components, subordinate-level units, and low-density/high-demand assets. Tasks are implied actions describing the accomplishment of desired effects or the avoidance of undesired effects. Figure 16 highlights this intrinsic interrelationship and the inherent links among end state, objectives, effects, and tasks. This is not the sole means for ei-

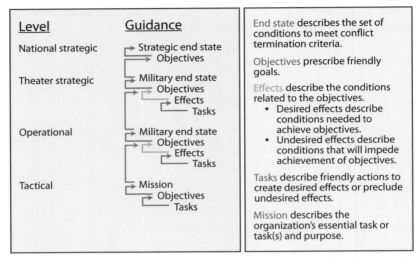

Figure 16. Relationships among end state, objectives, effects, and tasks

ther qualitatively or empirically measuring operational progress, but it represents a crucial assessment foundation for JFCs.

The concept of linking end state, objectives, effects, and tasks is far from a novel approach. History is full of examples. One of the most sophisticated illustrations occurred during the Mongol conquest of Vietnam in 1287–88. During the thirteenth century, the Mongol sphere of influence extended from China to Eastern Europe, and very few nations were capable of withstanding the Mongols' power. In 1287 they invaded Vietnam to expand their regional influence. When the Vietnamese were confronted by the Mongol onslaught, their national strategic end-state conditions were simple: regime survival and withdrawal of the Mongol forces. To achieve these end-state conditions, the Vietnamese, led by Tran Hung Dao, adopted a deliberate strategy of strategic withdrawal. The objectives were to extend the Mongol lines of communication beyond their capability to sustain their force and to lure the Mongols into terrain where they could not use their dominant force, their cavalry. The desired effect was the withdrawal of the invading Mongol forces. Another closely related Vietnamese objective was denial of Mongol freedom of movement on the land surface. The Vietnamese accomplished this by tasking their forces to destroy prominent bridges and interdicting key road networks. The cumulative effect created by this combination of objectives, effects, and tasks was to force the Mongols to withdraw along the river system. Using Ngo Quyen's victory over the Chinese in 938 as an example, Tran Hung Dao calculated that these actions would cause the Mongols to withdraw their forces via the Bach Dang River system (fig. 17).[11]

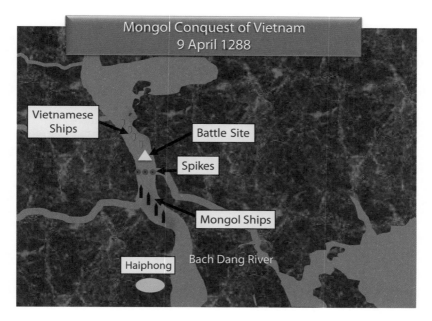

Figure 17. Battle of the Bach Dang River

After studying tidal projections, Tran directed that sharpened poles be cut, tipped with iron, and embedded in the river just to the north of modern-day Haiphong. On 9 April 1288, a small flotilla of Vietnamese shallow-draft boats engaged the Mongol fleet at high tide. When the Mongol fleet deployed for decisive action, the Vietnamese feigned retreat upriver and drew the Mongols after them in hot pursuit. As the tide fell, thousands of Vietnamese boats attacked the Mongol fleet at a prearranged point. When the Mongol fleet attempted to retreat to the sea, it was hopelessly caught in Tran's barrier system of iron-tipped poles, losing an estimated 400 ships and 6,000 men.[12] The Vietnamese ability to link effects to objectives and desired end-state conditions had a significant impact on their successful resistance to Mongol domination.

Aligning the End State, Objectives, and Effects with Centers of Gravity

An operation's end state, objectives, and effects have limited viability until they are aligned with a detailed systems analysis of the enemy and friendly centers of gravity. Joint doctrine defines *center of gravity* (COG) as "the source of power that provides moral or physical strength, freedom of action, or will to act."[13] COGs are often described using the Clausewitzian notion "hub of all power." From this perspective, if you destroy the COG, you cause the entire system to collapse. However, this is rarely the case because COGs are almost always protected and our adversaries are complex adaptive systems. What our doctrine does not clearly state is the purpose of a COG. The principal purpose of a COG is focus. Identifying COGs at the three levels of war establishes a clear delineation of priorities and responsibilities. It also produces a congruency mechanism that links actions at all three levels of war. The COG focus at the strategic level of war should create a directional foundation for the operational level of war, which should do the same for the tactical level of war. Assessing the dynamic characteristics of a COG bonds the desired end state, objectives, and effects with the strategy designed to defeat, destroy, neutralize, or protect a COG. However, the characteristics of a COG are far from simple. Figure 18 indicates why COG identification is often an elusive challenge.

Identifying a COG

Identifying a COG requires three important analytical and iterative steps. The first step is identifying the existing COGs at each level of war. The second step is a critical-factor analysis of each of those COGs to identify the decisive points and lines of operation leading into the COG. The final step is an analysis of how operations may shape and transform the COGs.

Determining a COG is a key step omitted from doctrine, and it requires a disciplined methodology. Contemporary doctrine focuses on critical-factor analysis without elaborating on *how* a COG itself is identified. The danger of this approach is aligning the end state, objectives, and effects against the wrong COG. At a minimum, JFCs and their staffs should consider enemy and friendly actors, their interests

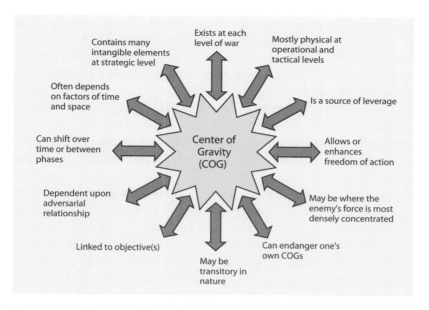

Exists at each level of war

Contains many intangible elements at strategic level

Mostly physical at operational and tactical levels

Often depends on factors of time and space

Is a source of leverage

Can shift over time or between phases

Center of Gravity (COG)

Allows or enhances freedom of action

Dependent upon adversarial relationship

May be where the enemy's force is most densely concentrated

Linked to objective(s)

Can endanger one's own COGs

May be transitory in nature

Figure 18. Characteristics of a center of gravity. (Reprinted from JP 5-0, *Joint Operation Planning*, 11 August 2011, III-23.)

and intent, key systems supporting those actors in the operational environment, and their strengths, weaknesses, and potential courses of action. This establishes a deliberate foundation for subsequent critical-factor analysis and reduces the potential for focusing on preconceived notions. Table 2 illustrates one method of arranging data and analyzing key systems to identify potential COGs of different actors. This type of analysis can also facilitate the assessment of political assumptions concerning a nation's support for, neutrality to, or hostility to an

Table 2. Identifying centers of gravity

Actor	Objective	Systems Analysis	Strengths	Weak-nesses	Courses of Action
		Political			
		Military			
		Economic			
		Social			
		Infrastructure			
		Information			

operation. However, it is imperative to link all COG assessments with friendly objectives and desired end-state conditions. As objectives and end-state conditions change, there is a high probability that the adversary's COGs will change. Additionally, there is a possibility the friendly COG may change.

It is also important to emphasize that identifying COGs requires a fundamental understanding of the three levels of war. Each level has specific functions, which should be examined to determine the desired effects at that level. The strategic level of war establishes national or multinational strategic security objectives and guidance and develops and uses national resources to achieve these objectives. Activities at this level include establishing national and multinational military objectives, sequencing initiatives, defining limits and assessing risks for the use of military and other instruments of national power, and providing military forces and other capabilities in accordance with strategic plans.[14] The operational level links the employment of tactical forces to achieving the strategic end state. At the operational level, commanders conduct campaigns and major operations to establish conditions that define the end state.[15] The tactical level involves the employment and ordered arrangement of forces in relation to each other.[16] Comprehending the functions of each level of war helps JFCs and their staffs identify the correct COGs and provide the right focus.

Conducting Critical-Factor Analysis

As figure 19 indicates, a COG is almost never a single node. COGs usually consist of multiple nodes with interconnecting relationships to the operational environment's systemic architecture.

The critical-factor analysis model (fig. 20) developed by Joe Strange and amplified in Jack Kem's *Campaign Planning: Tools of the Trade* provides an analytical tool for assessing a COG's key nodes.[17] This model examines three factors related to a COG: critical capabilities, critical requirements, and critical vulnerabilities.

Critical capabilities are crucial enablers for a COG to function and are essential to accomplishing an adversary's objectives. Critical requirements are the conditions, resources, and means that enable a critical capability to become fully operational. Critical vulnerabilities are those aspects or components of critical requirements that are de-

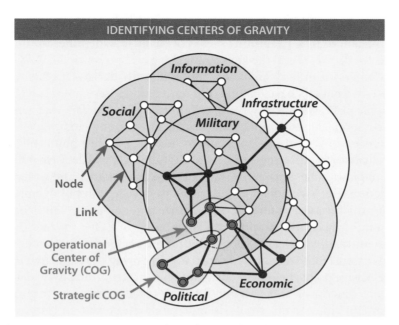

Figure 19. Systems perspective of operational environment. (JP 5-0, *Joint Operation Planning*, 26 December 2006, IV-11.)

Center of Gravity	Critical Capability
Source of power that provides freedom of action, physical strength, and will to fight	Means that are considered crucial enablers for the adversary's COG to function and essential to the accomplishment of the adversary's assumed objective(s)
Critical Vulnerability	**Critical Requirement**
Aspects or components of the adversary's critical requirements which are deficient or vulnerable to direct or indirect attack that will create decisive or significant effects disproportionate to the military resources applied	Those essential conditions, resources, and means for a critical capability to be fully operational

End State / Objective:

Figure 20. Critical-factor analysis model

ficient or vulnerable to direct or indirect attack achieving decisive or significant results. Examining these critical factors identifies the decisive points that reveal the keys to attacking and preserving COGs.

Centers of gravity are often confused with decisive points, but these two entities have distinct differences. A decisive point is a geographic place, specific key event, critical factor, or function that, when acted upon, allows commanders to gain a marked advantage over an adversary or contribute materially to success. Decisive points offer a mechanism for affecting a protected COG. COGs and decisive points should always be considered in relation to one another and never in isolation. Decisive points are also used to develop lines of operation that provide a vision for how to organize and employ US and coalition military efforts. Additionally, an analysis of COGs enables JFCs to identify the physical or geographic points, major events, functions, and systems that will ensure unified action during the campaign.

Decisive points belong to one of four basic categories: physical, key events, functional, and systemic. Examples of physical decisive points are major cities, rivers, straits, islands, command posts, and airspace. Key events include initial lodgment of friendly forces, culmination of the enemy's strategic reserves, establishment of bridgeheads, and elections. Functional and systemic decisive points are closely related to key events because they are either intangible or multifaceted, but they are different because of their purpose. Functional decisive points correspond to specific tasks or functions. Illustrations include establishing early warning; gaining air superiority; conducting reception, staging, onward movement, and integration (RSOI); and protecting the force. Systemic decisive points affect a node or combination of nodes within a system. Examples include a power-generation plant controlling an electrical grid, a fiber-optic link enabling telecommunications, and computer servers facilitating critical aspects of financial transactions. Analyzing decisive points assists JFCs and their staffs in determining and prioritizing the best methodology for affecting an adversary's COGs and seizing the initiative.

Operation Desert Storm is a classic case in point illustrating the connectivity between COGs and decisive points. During the construct of the operational design for Operation Desert Storm, the coalition identified the Iraqi command and control system as the operational COG. One of the decisive points protecting this COG was the Iraqi Kari system, an integrated air-defense command and control (C2) system that provided tracking and targeting information for Iraqi fighter aircraft and surface-to-air missiles (fig. 21). The Kari system included French and Italian long-range and short-range radars capable of detecting aircraft flying as low as 50 feet. Additionally, the system was augmented with a Japanese RM-835 ground-based system capable of tracking electronic emissions and Chinese Nanjing low-frequency radars.[18] This collective arrangement of equipment gave the Iraqis a dynamic early warning system capable of identifying the launch of air operations against their strategic and operational COGs.

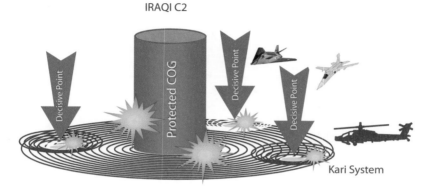

Figure 21. Relationship between the Iraqi COG at the operational level of war and the Kari system decisive point during Operation Desert Storm

Although no doctrinal critical-factor analysis was conducted during the planning for Operation Desert Storm, analysis of the Kari system revealed several key vulnerabilities. The system, controlled by four sector operation centers, consisted of a number of diverse radar systems with limited repair parts. Additionally, the system's 500 radars were located at approximately 100 geographically dispersed sites with fragile communications links and limited power.

A contemporary critical-factor analysis of the Kari system (fig. 22) shows the relationships of the Iraqi command and control COG to its critical capabilities, critical requirements, and critical vulnerabilities. It also indicates how analysts assessed critical vulnerabilities to determine where to attack. On 16 January 1991, the coalition selected and destroyed two radar clusters separated by 60 miles of open desert, opening the way for coalition aircraft to attack other decisive points related to Iraqi command and control.

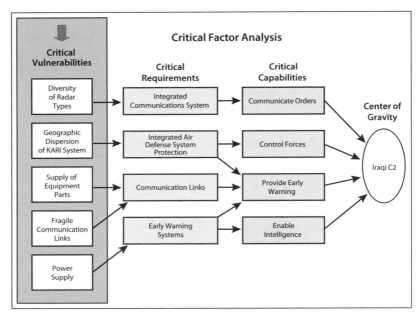

Figure 22. Critical-factor analysis of Iraqi command and control and the Kari system

It is important to emphasize that operational design requires planners to identify both friendly and enemy COGs and friendly and enemy decisive points. Any design that does not assess friendly COGs and decisive points creates a serious flaw in the operation. Analyzing friendly COGs and decisive points identifies what the friendly forces must protect and allows the friendly forces to prioritize defensive operations and the use of scarce resources.

Desert Storm offers another illustration of linking decisive points to centers of gravity. During Desert Storm, Iraq identified the coalition's strategic COGs as the coalition itself. In an effort to destroy the coalition, the Iraqi military fired Scud missiles into the state of Israel hoping to bring Israel into the conflict and disintegrate critical Arab support for the coalition. As a result of Iraq's strategy, coalition forces had to position Patriot air defense systems in Israel and modify air operations to protect Israel (fig. 23).

As JFCs and their staffs evaluate decisive points, they determine the most important ones and designate them as decisive points for

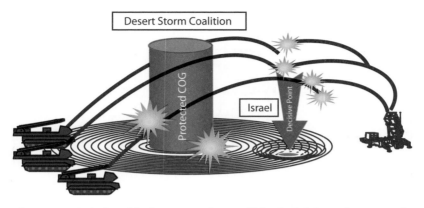

Figure 23. Relationship between the coalition's COG at the strategic level of war and the decisive point of Israel during Operation Desert Storm

the campaign. These designated decisive points become the basis for developing lines of operation (LOO), visualizations of a campaign's concept of operations that link tactical and operational objectives to the end state.[19] LOOs define the orientation of the force in time and space or purpose in relation to an adversary or objective. LOOs also assist JFCs in synchronizing military activities directed at a series of strategic and operational objectives to attain the military end state.

An operation's LOO can be described as physical, logical, or both. JP 5-0 defines a physical LOO as "a physical line that defines the interior or exterior orientation of the force in relation to the enemy or that connects actions on nodes and decisive points related in time and space to an objective(s)."[20] An example of a physical LOO is General MacArthur's island-hopping campaign in the Southwest Pacific during World War II. MacArthur maneuvered his forces along a geographic line of operation from Australia to New Guinea and then to the Philippines (fig. 24).

A logical LOO is "a logical line that connects actions on nodes and decisive points related in time and purpose with an objective(s)."[21] A logical LOO focuses on depicting an arrangement of tasks, effects, and objectives. The Allied liberation of Western Europe provides an excellent illustration of a logical LOO. On 12 February 1944, the Allied combined chiefs of staff gave General Eisenhower a one-page directive ordering the reconquest of Western Europe. Using the Allied

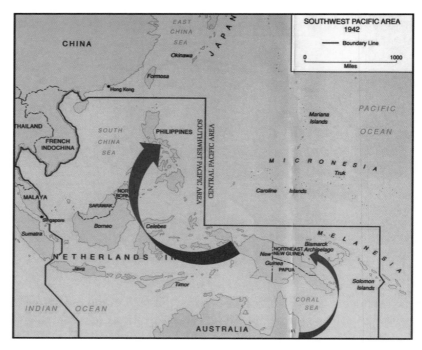

Figure 24. Southwest Pacific in World War II: physical line of operation. (Reprinted from Charles R. Anderson, *Papua Campaign Brochure*, US Army Center of Military History, n.d.)

combined chiefs of staff's guidance, General Eisenhower and his staff developed seven key decisive points aligned along a logical LOO for the campaign (fig. 25). Those decisive points were (1) establish a lodgment on the coast of Normandy, (2) conduct a breakout and build up a logistics base, (3) pursue the enemy on a broad front with three Army groups, (4) secure ports in Belgium and Brittany, (5) build up logistics in the vicinity of the Rhine, (6) complete the destruction of the enemy forces west of the Rhine, and (7) launch the final attack as a double envelopment.[22] What is important to note is that this plan's execution did not deviate significantly from its original intent.

The visualization of a concept of operations normally has multiple interconnected LOOs reflecting the simultaneous accomplishment of key tasks and objectives. The interconnectivity between lines of operation can also be used to show key decision points that connect the concept of operation with branches and sequels.

Figure 25. Operation Overlord: decisive points along logical lines of operation

How Operations Shape and Transform COGs

Planners must remain cognizant of the fact that COGs are dynamic. They can change based on the success or failure of military operations, changes in national policy, modifications to objectives, and enemy actions. As an example, a JFC designates an adversary's operational reserve as the COG at the operational level of war. His intent is to focus coalition forces on bringing the enemy to the culmination point, depriving him of attaining his military objectives. However, once coalition forces destroy the exploitation force, what is the new COG at the operational level of war? The reason for raising this issue is planners must think ahead to ensure that the relationships among LOOs, decisive points, COGs, objectives, effects, and the end state are properly aligned. Additionally, the identification of COG transformations assists commanders in setting the preparatory conditions necessary for executing the next phase of the operation. This does not mean that planners should approach COGs from a purely linear perspective. As figure 26 implies, the focus directed at COGs can overlap phases of an operation with components being directed to conduct asymmetric attack and establish the conditions required

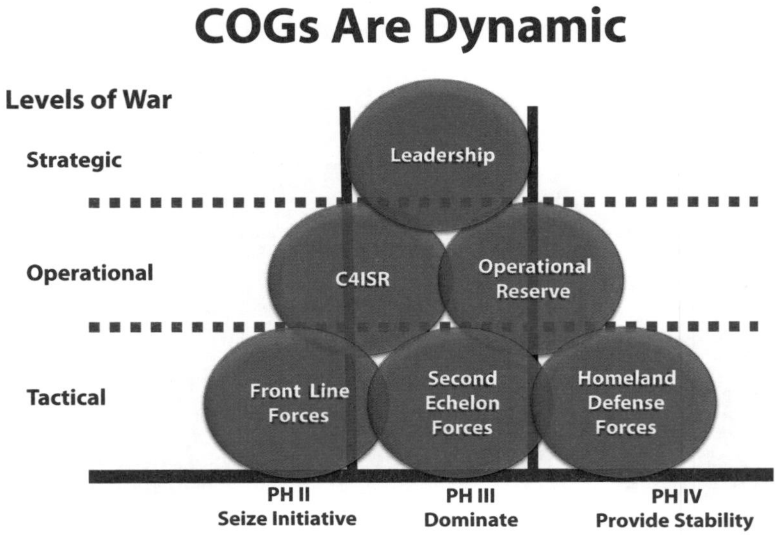

COGs Are Dynamic

Levels of War

Figure 26. Transformation of COGs as an operation's objectives change

for future phases. In general, COGs at the strategic level of war will rarely change as an operation progresses. At the operational level of war, COGs may change slightly, and at the tactical level they will most probably change.

It is also important to understand that the relationships among the levels of war change based on national strategic objectives and the complexities of the operational environment. If attainment of national strategic objectives requires conventional military operations, then the relationships among the three levels of war will more than likely be arranged in a hierarchical fashion. However, if military forces are employed in irregular warfare operations, the relationships among the levels of war are much more dynamic. In irregular warfare, seemingly insignificant actions at the tactical level of war can have immediate repercussions on the operational and strategic levels of war. Figure 27 shows the contrast in how different types of operations shape and transform the relationships among the levels of war.

The Arrangement of Operations

The arrangement of operations to accomplish military objectives and national strategic end-state conditions is one of the most impor-

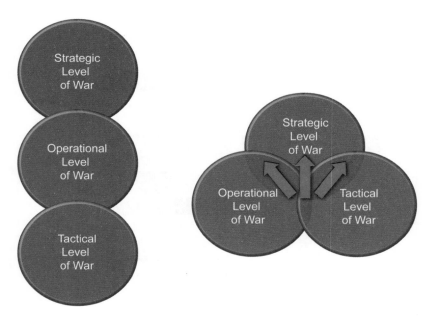

Figure 27. Transformation of the relationships among the levels of war

tant decisions a JFC will make. This decision involves a detailed consideration of how LOOs align with friendly and enemy COGs and the
vertical and horizontal relationship of decisive points between LOOs.
Figure 28 depicts the multifarious nature of these relationships and
the challenges in determining how best to arrange operations.

Commanders assess a host of essential factors when determining
how to arrange and link related military operations. These factors include the geography of the joint operation area, available strategic lift,
changes in command structure, force protection, logistics, adversary
reinforcement capabilities, and public opinion.[23] The doctrinal tool
for arranging the sequential and simultaneous nature of operations is
phasing. JP 5-0 defines phases as "definitive stage[s] of an operation
or campaign during which a large portion of the forces and capabilities are involved in similar or mutually supporting activities for a
common purpose."[24]

Phasing helps JFCs and their subordinates to visualize how an entire
operation will unfold and to determine force, resource, and time requirements. The principal benefit of a phase is that it assists commanders in achieving major objectives by planning manageable subordinate
operations to gain progressive advantages. Phases can be sequential or

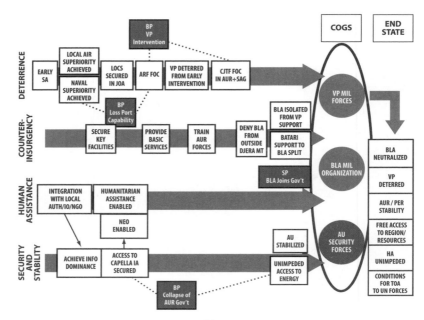

Key to abbreviations

ARF	Alliance Reaction Force	JOA	joint operations area
AU	Aurigan	LOCs	lines of communications
AUR	country of Auriga	MIL	military
AUTH	authority	MT	mountain
BLA	Batari Liberation Army	NEO	noncombatant evacuation operation
BP	branch plan	NGO	nongovernmental organization
CJTF	combined joint task force	PER	country of Perseus
COGs	centers of gravity	SA	situational awareness
FOC	full operational capability	SAG	country of Sagitta
HA	humanitarian assistance	SP	sequal plan
IA	international airport	TOA	transfer of authority
INFO	information	UN	United Nations
IO	information operations	VP	country of Vulpecula

Figure 28. The multifarious nature of arranging operations

simultaneous, and they can overlap. A transition from one phase to another, say from defensive to offensive operations, signals a change in emphasis. During planning, commanders establish conditions that should be met before transitioning to or initiating a new phase.

Joint doctrine provides extensive information on phases, but it does not state how phases are structured. Phases consist of six key entities: objectives, effects, start and end times, major tasks, priority of effort, and the desired phase transition or initiation criteria.

A template of this phase structure is as follows:

Phase I - Deter

Objectives: Deter aggression and prepare for hostilities.

Effects: Adversary's leadership recognizes the coalition's resolve and understands the risk of using military force; theaterwide force protection established in the event of attack.

Commences: Commences upon receipt of unambiguous warning.

Ends: Ends when hostilities are initiated.

Tasks: Monitor and confirm intelligence indicators; institute operational security directives; implement force protection measures; coordinate flexible deterrent options; flow force enhancements; employ counter special operations forces (SOF) operations; prepare reception, staging, onward movement, and integration assets; update pre–air tasking order.

Priority of Effort: First to air component command, followed in order by special operations component command, naval component command, and land component command.

Phase Transition or Initiation Criteria: Force protection established; command, control, communications, computers, and intelligence (C4I) protected; aerial and sea ports of debarkation prepared to receive follow-on forces; sea lines of communication secured; command prepared for hostilities.

Upon completion of each phase, planners must meticulously ensure that their operational design integrates all four of the national instruments of power. If planners focus solely on the military instrument of power, they lose control of the operational design by not leveraging the other three instruments of power. All diplomatic, informational, military, and economic courses of action should be forged into a single unified course of action. Additionally, the operational design should carefully examine potential first-, second-, and third-order effects of all military and nonmilitary actions.

For instance, if the strategic objective is to deter aggression, the operational design should incorporate other aspects such as an operational assessment of possible UN resolutions, the role of the media, the freezing of assets, and the imposition of economic sanctions. Failing to synchronize or deconflict nonmilitary initiatives can have severe and unintended consequences. As an illustration, if the Department of State were to implement a unilateral or uncoordinated plan to impose sanctions on a country such as North Korea, it could well provoke hostilities. This could produce cataclysmic results, especially if the theater headquarters were not prepared to respond.

Assumptions: The Forgotten Element in Operational Design

Joint doctrine defines an assumption as "a supposition on the current situation or a presupposition on the future course of events, either or both assumed to be true in the absence of positive proof, necessary to enable the commander in the process of planning to complete an estimate of the situation and make a decision on the course of action."[25] This complex definition, however, obscures the critical role that assumptions play in operational design. As a result, planners often see the primary role of assumptions as simply a tool to continue planning and not as a key framework element in operational design. *Assumptions endow operational design with the ability to identify the greatest risk to an operation.* This crucial ability assists JFCs and their staffs with determining branches and sequels, pinpointing decision points, and developing the initial commander's critical information requirements (CCIR).

Failure to properly assess assumptions can have catastrophic and historic consequences, as it did at the battle of the Teutoburger Wald in 9 AD.[26] This obscure battle changed Rome's vision of hegemony and quite possibly history. It began as a campaign to subdue rebellious Germanic tribesmen, and it ended in the halting of Roman expansion beyond the Rhine River. At the time of this battle, the Romans were the masters of operational design. They had superior organization, training, and advanced technology and were confronting an adversary they had previously slaughtered like cattle. However, they failed to identify two critical assumptions. The first assumption was that the Germanic tribesmen serving in the Roman army would remain loyal allies. The second was that the Roman legions would be able to deploy their forces and not have to fight on the march in the forests.

In the battle, an alliance of Germanic tribes under the military leadership of Arminius defeated the XVII, XVIII, and XIX Roman legions under Quinctilius Varus. According to the Roman historian Gaius Cornelius Tacitus, the Roman legions, composed of more than 20,000 men, were ambushed in the forests of the Teutoburger Wald and annihilated by Arminius's warriors. Those who survived were crucified, buried alive, or sacrificed to the Germanic gods. The battle is significant because it marked the end of Rome's expansion into the Germanic frontier.[27] Additionally, this defeat shows that Roman over-confidence trivialized its assessment of assumptions.

In many ways the assumptions derived during operational design define the overall quality of an operation plan (OPLAN). If planners approach an OPLAN's assumptions as just another mechanical step in the JOPP, the probability of a plan surviving first contact decreases substantially. Assumptions are not merely a list in the OPLAN for enhancing the plausibility of the operational design. They require constant vigilance, and they should neither be forgotten nor discarded because they identify the greatest risk to the successful execution of the OPLAN. Consequently, in case an assumption proves not to be true, a corresponding branch or sequel should be developed in advance to prevent disaster and ensure success of the campaign. This emphasis on assumptions in turn creates a foundation for establishing CCIRs that assist the JFC in making effective decisions during execution. It does this by focusing CCIRs toward the most critical decisions a JFC must make. Of course if an assumption is invalidated, any decisions based on that assumption should be immediately reexamined.[28]

Another indicator of the quality of an OPLAN's operational design is the correlation of assumptions to the levels of war. As planners develop assumptions, they should align those assumptions with the appropriate level of war capable of planning and executing the branch or sequel. For example, consider an assumption concerning the availability of strategic lift for the initial phases of a campaign. An operational-level headquarters has almost no ability to influence the development of a branch or sequel involving strategic lift because it does not control the assets. These assets are controlled at the strategic level of war, and any assumptions concerning them should be incorporated into the strategic guidance and planning. This aligns responsibility for planning any associated branches or sequels for strategic lift with the level of war most capable of executing the branch or sequel. It also keeps the operational level of war from planning a branch or sequel it cannot execute. This, however, does not mean that the operational-level headquarters can ignore assumptions made at the strategic level. JFCs and their staffs must still plan appropriate responses, but their responses to strategic-level assumptions are tailored to planning areas they can affect.

Assumptions have three basic characteristics. They should be logical, realistic, and essential for the planning to continue.[29] There are no "cookie cutters" for determining assumptions. But there are some common categories of assumptions to consider in operational design. These categories are time, politics, forces, and the enemy.[30]

At the operational level of war, time is arguably the most important assumption, especially for power-projection nations. Time drives the ability to prepare, deploy, and generate forces. It also influences the effectiveness of execution, cost, and everything related to an operation. If an operation assumes C = D based on an unambiguous warning, then deployment and the operation begin simultaneously. (C-day is the unnamed day on which deployment for an operation begins or is to begin. D-day is the unnamed day on which a particular operation begins or is to begin.) If C = D - 5, then deployment begins five days prior to the beginning of the operation. If either of these assumptions proves to be false, then a branch or sequel must be available for execution.

Political assumptions encompass a wide range of issues such as access to host nation support, basing rights, overflight routes, and nation neutrality. As with all assumptions, political assumptions are inherently complex and must be constantly assessed for changes through focused CCIRs. An example is the assumption that China would not commit military forces to combat during the Korean War. On 7 October 1950, the UN General Assembly approved a British-sponsored resolution authorizing UN Command (UNC) forces to occupy all of North Korea. This resolution was based on the belief that China would not commit combat forces to support the North Koreans. At the time, Chinese communist forces (CCF) were still battling 400,000 Nationalist Chinese forces for control of China. The Chinese economy was in shambles, and the CCF were underequipped and undertrained. In spite of these challenges, on 19 October 1950 China introduced an initial Chinese People's Volunteer Force (CPVF) estimated at 260,000 men. The CPVF fought throughout the duration of the Korean War and suffered over 539,000 casualties.[31] It is interesting to note that China may have decided as early as August 1950 to intervene in Korea, but there was little if any attempt to collect and evaluate CCIRs against this assumption.[32]

Assumptions about enemy forces are critical also, and it is vitally important to never "assume away" an enemy capability. Examples of enemy assumptions include where the enemy's main effort will be, when the enemy's activities will culminate, and whether the enemy will use weapons of mass destruction. Historically, many plans have failed because they underestimated an enemy's capability and did not have a requisite branch or sequel prepared to counter that capability. Operation Iraqi Freedom (OIF) gives us an excellent example. During the planning for OIF a central assumption was that regular Iraqi army units would "capitulate and provide internal security."[33] This assumption was a key element in the decision to limit the amount of combat power deployed to Iraq, and it had a significant impact on the effectiveness of stability operations when this proved to be false.

Forces are another category of assumptions vital to power-projection nations because they have significant and complex interrelationships with all categories of assumptions. If a power-projection nation misjudges an assumption about forces, dire consequences can follow. Most assumptions in this category revolve around the availability of forces for employment. These include not only US forces, but also such assets as prepositioned war stocks, maritime prepositioned sets afloat, and coalition forces.

It is also important to understand that assumptions can have cascading effects on decisions made about forces. OIF is again an excellent illustration. The decision that OIF's force structure was sufficient to accomplish the strategic end state proved to be severely flawed for several interconnected reasons. First, this decision was directly linked to three critical planning assumptions:

1) The Iraqi regular army would "capitulate and provide internal security."

2) Iraqi resistance was unlikely.

3) Postwar Iraq would not be a US military responsibility.

Second, on 23 May 2003 the Coalition Provisional Authority dissolved the Iraqi Army, which the US Central Command (USCENTCOM) commander had assumed would provide internal security. Third, a Government Accountability Office (GAO) analysis observed that the OIF OPLAN did not document risk mitigation strategies in case assumptions were proven wrong. According to a 2006 report by the Joint Center for Operational Analysis, OIF planning did not examine the consequences of assumptions, which contributed to the inability of coalition forces to prevent the breakdown of civil order in Iraq.[34]

OIF highlights the direct correlation among assumptions, operational design's planning framework, and decision analysis. Proper development of assumptions allows JFCs and their staffs to frame risk and determine risk levels. Assumptions also identify major decision points for JFCs. Planners must make every effort possible to validate or invalidate assumptions prior to reaching critical decision points.

Notes

1. Joint Publication (JP) 1-02, *Department of Defense (DOD) Dictionary*, 126.
2. US Army War College, *Campaign Planning Primer AY 07*, 10.
3. US Department of Defense, *2008 National Defense Strategy*, 6.
4. Bush, *President's Address to a Joint Session of Congress and the American People*.
5. US Department of Defense press briefing, Rumsfeld and Myers.
6. National Security Directive 54, *Responding to Iraqi Aggression*, 2.
7. JP 5-0, *Joint Operation Planning*, 11 August 2011, III-19.
8. JP 1-02, *DOD Dictionary*, 367.
9. JP 5-0, *Joint Operation Planning*, 26 December 2006, III-9.
10. JP 1-02, *DOD Dictionary*, 265.
11. Grant, ed., *1001 Battles That Changed the Course of World History*, 181–86.
12. Chapius, *A History of Vietnam from Hong Bang to Tu Duc*, 84.
13. JP 3-0, *Joint Operations*, 11 August 2011, GL-6.
14. JP 5-0, *Joint Operation Planning*, 26 December 2006, GL-22.
15. Ibid., GL-19.
16. Kem, *Campaign Planning*.
17. Ibid. and Strange, *Centers of Gravity and Critical Vulnerabilities*.
18. Gordon and Trainor, *The Generals' War*, 105.
19. US Army War College, *Campaign Planning Handbook*, 93.
20. JP 5-0, *Joint Operation Planning*, 26 December 2006, IV-20.
21. Ibid.
22. D'Amura, "Campaigns," 42–51. See also Pogue, *Supreme Command*, which provides a description of General Eisenhower's wartime command, focusing on the general, his staff, and his superiors in London and Washington and contrasting Allied and enemy command organizations.
23. JP 5-0, *Joint Operation Planning*, 11 August 2011, III-35.
24. Ibid., IV-33.
25. JP 1-02, *DOD Dictionary*, 30.
26. Fuller, *Military History of the Western World*, 239.
27. Wells, *The Battle That Stopped Rome*.
28. US Army War College, *Campaign Planning Handbook*, 38.
29. JP 5-0, *Joint Operation Planning*, 11 August 2011, IV-8.
30. US Army War College, *Campaign Planning Primer*, 17.
31. Li, Millett, and Yu, *Mao's Generals Remember*, 6.
32. Jian, "Sino-Soviet Alliance," 26.
33. US Army, *On Point*.
34. US Joint Forces Command, *Operation Iraqi Freedom May 2003 to June 2004*.

Chapter 4

Establishing the Link between Operational Design and Decision Analysis

In recent years numerous publications have expanded the exploration of operational design beyond our current joint doctrine. Their focus has been how to integrate operational design into the JOPP and what role commanders play in the process. As a result, operational design is evolving as a leader-driven process oriented on problem framing. The intellectual process of problem framing is extremely valuable for developing a solution; however, deriving a correct solution does not necessarily guarantee success. To have success, the commander must apply decision analysis to implement the solution. The difference between problem framing and decision analysis is that decision analysis produces decision criteria to aid a commander in making the best decision.

The structure that operational design provides shapes decision analysis directly and indirectly throughout the JOPP. However, there are three critical junctures in the JOPP where operational design has the potential to significantly influence the effectiveness of decision analysis. The first juncture is in the JFC's overarching vision for the theater or operational strategy. The second is during course analysis and war gaming, and the third is just after the selection of the accepted course of action. Figure 29 illustrates these junctures and the commander/staff operational design interaction during the JOPP.

As an example of design's link to decision analysis, if the forward edge of the battle area (FEBA) is penetrated by a major force, the correct decision may be to commit the operational reserve. But if the reserve is committed too soon or too late—or along the wrong axis—it may result in a catastrophic failure.

The cognitive map's holistic view, combined with its capacity to identify decision points, is one of the principal reasons why operational design should be a commander-driven process that integrates into the JOPP. In contrast, the JOPP is primarily a staff-driven process that takes its direction from the JFC's vision embedded in the command's operational design. Operational design should begin at the initiation of the JOPP and continue through to the completion of the plan or the execution of the order. The primary reason for this is

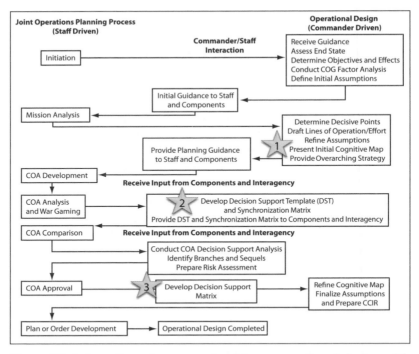

Figure 29. Integration of commander-driven operational design with the joint operation planning process

that operational design should set the foundation for the analysis of preplanned and emergent-opportunity decisions.

Overarching Vision for Strategy

As the JFC completes the initiation and mission-analysis steps of the JOPP, one of the products he should develop is a preliminary vision of the theater or operational strategy. Strategy has been characterized by terms such as *annihilation, attrition, deterrence, decapitation, disaggregation*, and *diffusion*. JP 1-02 defines theater strategy as "concepts and courses of action directed toward securing the objectives of national and multinational policies and strategies through the synchronized and integrated employment of military forces and other instruments of national power."[1] The vision the JFC constructs functions as a guideline for COA development. It should be as generic as possible so that it does not prejudice courses of action, but it should also have enough structure to ensure that the staff understands what the commander wants. One methodology embedded in operational design for achieving this is the cognitive map described in the previous chapter.

A cognitive map reflects the intrinsic relationships among LOOs, decisive points, centers of gravity, objectives, effects, and the end-state conditions. This visualization assists JFCs with determining the arrangement of multiple LOOs and assessing the risks associated with assumptions. But most importantly, the cognitive map identifies initial decision points critical to the success of the operation. Note that *decision* points are different from *decisive* points. A decision point is a position in space and time when the commander or staff anticipates making a key decision concerning a specific course of action.

Figure 30 depicts a JFC's initial cognitive map developed prior to providing COA guidance. The map indicates decisive points, LOOs, enemy COGs, objectives, and end-state conditions. The map also denotes two key decision points that will shape the strategy. The first decision point on the deter LOO identifies the crucial point where deterrence fails. This decision point will mandate a military recom-

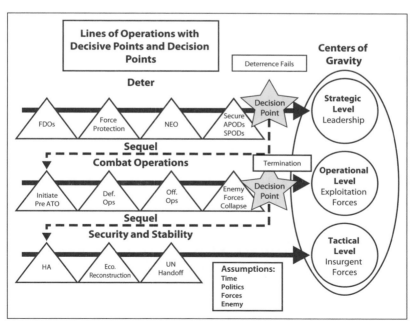

Figure 30. Cognitive map outlining an initial strategy

mendation and a political decision to use force. The staff should take this guidance and incorporate it into courses of action being developed. The second decision point on the combat operations LOO identifies a decision to recommend termination of combat operations. In this case, planners assess that the enemy has limited strategic objectives and will sue for peace prior to friendly forces achieving all military termination criteria. This preplanned decision point allows the JFC to formulate recommended decision criteria for political leaders and determine how it affects the theater strategy and the attainment of national or multinational strategic objectives. The main point of this example is that developing a sound operational design enhances the JFC's ability to envision the best strategy for the operation or campaign. This strategy, linking the application of military force to the attainment of national or multinational strategic objectives, forms the basis for COA development.

When JFCs incorporate their strategic vision into COA development, the staff's productivity increases significantly. This also prevents the myopic exploration of divergent (e.g., preemptive and nonpreemptive) courses of action. Directing the development of a nonpreemptive course of action is largely an unproductive activity because of the exceptional risk to the military force involved. No one should expect a US or multinational military force to stand idly by while an adversary strikes a blow. The more astute guidance is to develop courses of action with branches and sequels that give political leaders the flexibility to use military force if deterrence fails. This type of guidance also forms a much better foundation for identifying governing factors to select the best course of action.

In operational design it is imperative to understand that a large number of factors shape theater strategy, but one of the most critical influences on strategy is power projection. The Mongol conquest of Transoxiana, modern day Uzbekistan, provides a relevant example for today. In 1218 a Mongol caravan with emissaries was transecting the territory of Transoxiana ruled by Shah Ala al-Din Muhammad II. One of the Shah's governors, Inalchuq, interdicted the caravan in the vicinity of Otrar and slaughtered its 450–500 occupants. This event occurred as the Mongols were heavily engaged in the conquest of north China. As a result, instead of applying military force Genghis Khan sent three emissaries to the Shah: two Mongols and a Muslim. The Shah shaved the two Mongols, beheaded the Muslim, and sent his head back to Genghis Khan with the two Mongol ambassadors. Outraged at this provocation, Genghis assembled 90,000 men and marched to Transoxiana.[2] The initial engagement occurred along the Syr Darya River in the Fergana Valley. The Shah's forces, with an estimated 400,000 men, defeated the Mongol attack. However, Genghis used this setback to fix and envelop the Shah's forces. The Mongols marched north and attacked the city of Otrar, capturing the governor, Inalchuq, who had attacked the Mongol caravan. The Mongols put Inalchuq to death by pouring molten silver into his eyes and ears and continued their attacks.[3] Stunned by the Mongol advances, the Shah's forces were paralyzed. The Mongols then attacked the strategic cities of Bukhara and Samarkand, causing the Shah's forces to completely collapse.

At first glance the salient lesson appears to be that a strategy of asymmetric attack caused catastrophic psychological paralysis. But the real lesson in the Mongol strategy is power projection. The Mongols were able to project a light, lethal force over 2,000 miles across the Tien Shan Mountains without the aid of a modern transportation system.

Today this lesson is just as important as it was in 1219–20. You can have an adaptive and sophisticated strategy, but before you can bring force to bear you must be able to get to the fight. This basic requirement places some significant limitations on strategy. As JFCs develop their strategy, they must judiciously assess the combat power needed to achieve the strategic political objectives and their ability to project that power. Assuming that there will be time to build a complete force is high risk. The international community has learned through Operations Desert Shield, Desert Storm, Allied Force, and Iraqi Freedom that if an opponent is afforded the freedom to build combat power, the consequences are grave. Strategic lift will limit the initial agility of strategy in most current operational environments. Therefore, JFCs are initially left with a baseline choice of a denial or halt strategy. A halt strategy focuses on simply stopping the enemy. In contrast, a denial strategy concentrates on bringing the adversary to culmination, preventing him from accomplishing his military and political objectives. Both of these approaches have a tremendous impact on COA development. Commanders cannot develop a cogent concept of operations or a scheme of maneuver without knowing how their force builds.

Based upon the JFC's strategy, power projection should build on four rudimentary and integrated force modules: command, control, communications, computers, and intelligence, surveillance, and reconnaissance (C4ISR); combat power; force protection; and sustainment. Each of these modules can be combined into joint and multinational force packages designed to accomplish specific objectives. The intent here is not to conduct a transportation feasibility analysis before developing and analyzing courses of action. Rather, the main point is for the JFC to have strategy guidance that incorporates a force flow vision. Figure 31 provides a conceptual overview of modularly arranging forces into force packages to accomplish specific deployment objectives. In this case, based on the operational environment, the JFC gives C4ISR the initial priority of force flow to build situational awareness. As the force flow progresses, the priority shifts to force generation, then interdiction, and finally to combat forces for maneuver. The force flow then continues in the form of additional packages until all of the JFC's forces have closed in theater. Arranging forces in modules and packages provides enhanced operational flexibility to the JFC.

Additionally, understanding force-generation priorities empowers a JFC's ability to maximize the effectiveness of a deterrence phase and, if required, have the force necessary to accomplish the national strategic objectives. The basic principles of the construct "strategic preclusion" are valid. JFCs should develop a strategy that employs joint maneuver and interdiction forces capable of moving with such speed and with such overmatching lethality that a potential enemy cannot "set" his forces and operate at an advantage.[4]

Power Projection and Theory of Strategic Preclusion

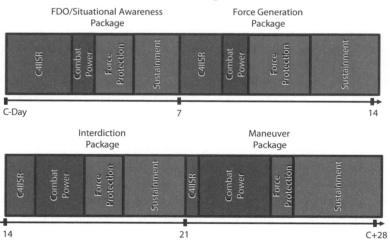

Figure 31. Notional arrangement of modules into force packages

Course of Action Analysis and War Gaming

The second juncture at which operational design's framework influences decision analysis is during the COA analysis and war-gaming step. Most of the war gaming conducted today is automated; however, manual war gaming provides JFCs the opportunity to evaluate courses of action against a complex, adaptive adversary. Manual war gaming is one of the most difficult activities in the JOPP. It requires a well-defined structure, discipline, extensive preparations, and rehearsals. In spite of these costs, war gaming provides an operational fidelity that is invaluable to the JFC. It sets the foundation for the development of the decision support template (DST) and the decision support matrix (DSM).

The structure of a war game is composed of the war-gaming technique, the war-gaming staff, and the physical layout of the war-gaming space. The selection of the war-gaming technique depends on the amount of time available to war-game and the JFC's focus. The two most commonly used techniques are war gaming by phase and by decisive points. The JFC and his staff can use one or a combination of the techniques. War gaming by phase is the most comprehensive approach and usually requires the most time (fig. 32). This technique assists the JFC with assessing specific phases or an entire operation.

War gaming by decisive points provides the JFC and his staff the ability to focus on a geographic area, key event, function, or critical factor during a specific period of time (fig. 33). Examples of such

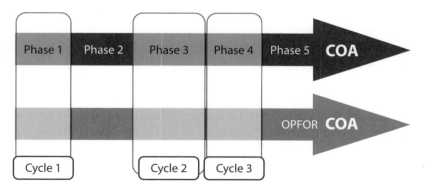

Figure 32. War gaming by phase. (Reprinted from Führungsakademie der Bundeswehr, *Wargaming: Guide to Preparation and Execution* [Hamburg, Germany: Führungsakademie der Bundeswehr, 2006], 8.)

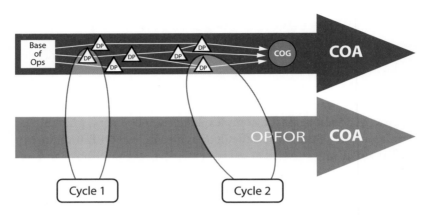

Figure 33. War gaming by decisive points. (Reprinted from Füh-rungsakademie der Bundeswehr, *Wargaming Guide to Preparation and Execution* [Hamburg, Germany: Führungsakademie der Bundeswehr, 2006], 8.)

decisive points include a constrained sea line of communication, lodgment operations, and key command and control transitions.

The staff organization supporting war gaming normally consists of four to five major sections: the executive section, support section, blue cell, white cell, and red cell (fig. 34). The executive section directs and controls the war game. The section's war-gaming director has overall responsibility for executing the JFC's guidance during the war game. The coordinator facilitates preparation and setup of the war game, and the referee adjudicates blue- and red-cell moves. The support section consists of the component liaison officers (LNO) and operational analysts, who provide detailed analysis of component contributions and joint functions. The blue cell consists of joint planning group (JPG) members responsible for developing, executing, and assessing the effectiveness of the friendly course of action being war-gamed. The white cell is optional. Its primary role is playing non-belligerent actors who may have a significant effect on a course of action. A good example of this is China's potential effect on courses of action involving the Korean peninsula. Each of these elements has crucial responsibilities in war gaming, but perhaps the most important cell impacting decision analysis is the red cell.

As a JPG forms, one of the JPG director's first tasks is establishing a dedicated red cell to analyze enemy courses of action. The red cell should do much more than simply plan red moves. It should prepare

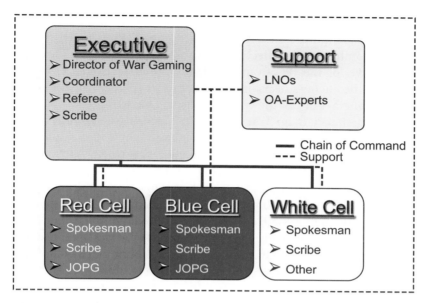

Figure 34. War-gaming sections. (Reprinted from Führungsakademie der Bundeswehr, *Wargaming Guide to Preparation and Execution* [Hamburg, Germany: Führungsakademie der Bundeswehr, 2006], 8.)

a red operational design focused at a minimum on the enemy's most likely and most dangerous courses of action. The operational design should identify enemy assumptions, branches, sequels, and associated decision points. Additionally, the red cell should analyze how red leaders collect intelligence, make decisions, and conduct their planning, decision, and execution cycle.

Understanding adversarial decision making is an absolutely essential planning element for JFCs because it links the application of military force to the attainment of national strategic end-state conditions. In simple terms, when friendly forces face an adversary, they are trying either to deter that adversary from action or compel him to take an action. The cumulative effects of inappropriately applied military force may move friendly forces further away from attaining the national strategic end-state rather than closer. To set the foundations for this analysis, the red cell should analyze the enemy's decision making model. The tendency for most westerners is to approach decision making from a rational model based on cost-benefit association. How state and nonstate actors make decisions, however, is far more complex than this, especially when one considers the effects of cultural, ideological, and religious dimensions. When the red cell has the ability to replicate the adversary's decision-making model, it substantially enhances the effectiveness of war gaming and builds a key perspective for the JFC to make friendly force decisions. Comprehending the enemy's decision analysis model will also yield indications of whether or not the enemy is altering his strategy.

During the Vietnam War the North Vietnamese skillfully alternated the emphasis of their strategy between conventional and unconventional operations (fig. 35). In 1965 the North Vietnamese Army (NVA) and National Liberation Front (NLF) attempted to use battalion-sized conventional attacks to achieve their objectives. As the Americans applied conventional forces to counter this maneuver, the NVA shifted to irregular warfare. Unable to accomplish its strategic objectives through irregular means, the NVA launched the Tet offensive of 1968. When this failed, it reverted back to irregular warfare. In 1972, as the Americans attempted to withdraw their forces under the policy of Vietnamization, the North Vietnamese launched the Nguyen Hue (Chiến dịch Xuân hè) or Eastertide offensive. This offensive cost the NVA over 125,000 lives; however, it set them in a

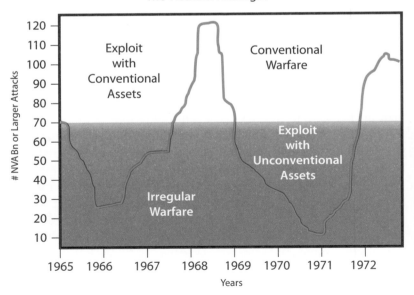

Predicting Effects
The Vietnam Paradigm

Figure 35. North Vietnamese alteration of strategy, 1965–72

strategic position to overrun the South Vietnamese government in 1975.[5] The focal point of this historical anecdote is not the alternation of strategy, but the use of the red cell to predict alternations.

All sections involved in war gaming must understand the sequence for the war game and approach it in a disciplined manner. Discipline emanates from two sources of the war-game structure. The first is the designation of the spokesmen for the red, blue, and white cells. Each cell should ideally have only one spokesman, who must be able to fully articulate the team's actions. This minimizes potentially distracting discussions that fail to contribute to the war-game process. The second source is time, and it serves as the principal means for instilling discipline in the war-game process. This element of structure keeps the war game from disintegrating into a series of peripheral discussions about a course of action. When the JPG decides to conduct a war game, it requires a tremendous investment in time and resources. If the JPG director allows the structure to disintegrate, the return on investment is significantly diminished. The intent of war gaming is threefold: to assist the JFC in selecting the best course of action, identify COA modifications, and conduct decision analysis. The war-gaming sequence, combined with strict time limitations, provides the vital element of discipline. A notional sequence for one war-game move is depicted in figure 36. The war-gaming sequence used most often is a basic action-reaction-counteraction move culminated by a cognition phase.

Prior to the first move, the war-gaming coordinator explains to all sides how the war game will be executed. Then the war-gaming director sets the context of the operational environment by reviewing the strategic setting, salient characteristics of the area of responsibility (AOR), the belligerents, and key events leading up to the start of the war game. With this foundation established, the blue and red cells preview the courses of action to be war-gamed, their order of battle, and the disposition of their forces. Based on this information, the war-gaming director decides which side has the initiative.

When the war-gaming cells begin the action step of the war game, the team with the initiative provides an overview of its activities during the designated operational time frame. The opposing team is then allowed to ask any questions for clarification, and the referee adjudicates any controversial activities identified during the sequence. When this concludes, the scribes from the executive section and red, blue, and white cells record the results.

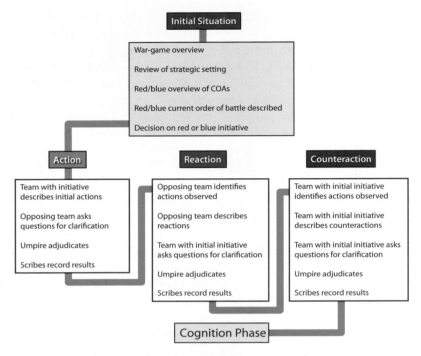

Figure 36. Action-reaction-counteraction war-gaming sequence

In the reaction step, the side without the initiative identifies the opposing side's actions it observed and summarizes its reactions. When it finishes, the opposing team asks questions for clarification, and the referee adjudicates any controversial activities. The scribes record the results and insights from this step. Figure 37 shows this sequence in a time line.

The side with the initiative then begins the counteraction step by describing what it observed during the reaction step and giving an overview of its counteractions. When this concludes, the side without the initiative asks questions for clarification, and the referee provides adjudication, if necessary. The scribes then record the results and insights from this step.

The most important step in a war-game move is the cognition phase. During this step the component LNOs, subject matter experts, and operational analysts are polled for their insights on the war-game move. The scribes review their conclusions for the plenary war-game group. The focal points for this analysis are decision points, decision

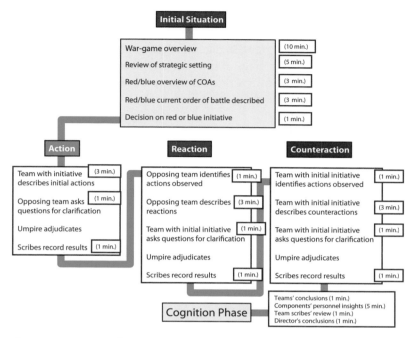

Figure 37. Notional war-gaming time line for one move

criteria, CCIRs, risk, and recommended modifications to the course of action. The war-gaming director then determines the conclusions from the move and directs that the appropriate changes be recorded in the COA decision support template.

A DST is a graphic record of the war game consisting of the concept of operations graphics and the synchronization matrix. It "depicts decision points, timelines associated with movement of forces and the flow of the operation, and other key items of information required to execute a specific friendly course of action."[6] The DST also identifies named areas of interest (NAI) and target areas of interest (TAI). NAIs are the geographical areas where "information that will satisfy a specific information requirement can be collected. Named areas of interest are usually selected to capture indications of adversary courses of action, but also may be related to conditions of the operational environment."[7] TAIs are "the geographical area[s] where high-value targets can be acquired and engaged by friendly forces. Not all target areas of interest will form part of the friendly course of action; only target areas of interest associated with high priority targets are of interest to the staff. These are identified during staff planning and war gaming. Target areas of interest differ from engagement areas in degree. Engagement areas plan for the use of all available weapons; target areas of interest might be engaged by a single weapon."[8] Figure 38 shows a decision support template combining decision graphics with a synchronization matrix.

Decision Support Template

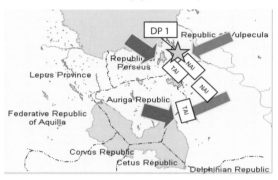

Synchronization Matrix

	Activity/Time	D-20	D-10	D-Day	D+1	D+2	D+3	D+4	D+5	D+10
	Phase	Deterrence			Seize Initiative				Dominate	
OPFOR										
Components	LCC									
	MCC									
	ACC									
	SOCC									
	POTF									
	JIACG									
Joint Functional Areas	C2									
	Intel									
	Sustainment									
	Info Ops									
	Civil Affairs									
	MILDEC									
Assessment	Decision Pts									
	Branch									
	Sequel									
	CCIR									
	Risks									

Figure 38. The DST composed of map graphic and synchronization matrix

Key to abbreviations in figures 38–42

AA	avenue of approach	JIACG	joint interagency coordination group
ACC	air component commander		
ATO	air tasking order	LCC	land component commander
AUR	country of Auriga	MCC	maritime component commander
Bde	brigade		
BLA	Batari Liberation Army	MEB	Marine expeditionary brigade
C2	command and control	MILDEC	military deception
C4ISR	command; control; communications; computers; and intelligence, surveillance, and reconnaissance	NAI	named area of interest
		OEF	operational exploitation force
		Op. res.	operational reserve
		OPFOR	opposing force
CCIR	commander's critical information requirement	Ops	operations
		PER	country of Perseus
CHOP	change of operational control	POTF	psychological operations task force
CJTF	combined joint task force		
COG	center of gravity	Pts	points
COIN	counterinsurgency	SOCC	special operations component commander
DP	decision point		
FEBA	forward edge of the battle area	SOF	special operations forces
FOM	freedom of movement	TAI	target area of interest
HA	humanitarian assistance	TBM	theater ballistic missile
Info	information	Vul	Vulpeculan
Intel	intelligence	WARNORD	warning order
IO	information operations	WMD	weapons of mass destruction
JFLCC	joint force land component commander		

Decision Support Matrix

The third juncture in the JOPP where operational design has the potential to enhance decision analysis is after the JFC approves the course of action. Once the JFC approves a course of action, it forms the basis for the plan's strategic concept, providing an opportunity to develop a decision support matrix and finalize assumptions. The collective analysis derived from the DST's record of enemy actions, friendly actions, NAIs, and TAIs establishes a crucial foundation for determining key decision points. JP 1-02 defines a decision point as "a point in space and time when the commander or staff anticipates making a key decision concerning a specific course of action."[9] However, identification of decision points is only half a DST's value. The other half is formulating the decision criteria for the JFC's preplanned decisions and emergent opportunities. This information is captured in a decision support matrix that records each key preplanned decision and decision criteria that evolve from the DST analysis and refinement of assumptions.

The following heuristic example links the war-gaming analysis captured in the DST (fig. 39) with the creation of a DSM. One of the most crucial decisions a JFC may have to make is the commitment of

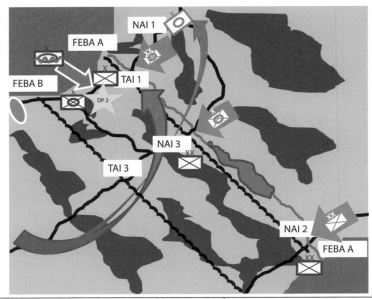

Decision Support Template
Phase II Seize Initiative

Commences: G-Day (the day on which an order is or is due to be given to deploy a unit)	**Tasks:** - Defend FEBA A - Establish local air superiority - Conduct special reconnaissance - Establish local sea superiority - Secure Capella International Airport (CIAP) - Generate force for dominate phase
Ends: - G+20 or culmination of enemy forces	
Objectives: - Deny enemy attainment of objectives - Set conditions for offensive operations - Stabilize Aurigan government	**Priority of Effort:** LCC, ACC, MCC, SOCC
Effects: - Force enemy to culminate forward of FEBA A - Interdict enemy exploitation forces - Delegitimize BLA insurgency	**Phase Transition Criteria:** Enemy forces culminated forward of FEBA A; conditions set for offensive operations

Time	G+1
ENEMY **Vul** **BLA**	Multidivision atk along coastal AAs in Perseus and Auriga; Sptd by insurgents Threatening Bde size penetration FEBA A; OEF moving toward coastal AA NAI 1 - Insurgents blocking LLOCs vic. Aurigan capital of Capella
CJTF	Main body in Sagitta. Coalition Response Force (CRF) BCT+ in Perseus. Priority of effort: Denial ops vic Perseus coastal AA
LCC	Perseus's forces at < 70%; tactical reserve committed
ACC	Local air superiority vic Perseus coastal AA Defensive counterair-APODS, Perseus SPODs Counterland (CAS/AI) – priority of effort Perseus Coastal AA Strategic attack of Vulpeculan WMD employment sites
MCC	Move SAG (MD) to vic Perseus Nth Coast SPODs; establish local sea superiority vicinity Perseus coast; conduct countermine ops; prepare to CHOP MEB to LCC
Intel	Priority of ISR to NAI 1 and TAI 1
IO	Highlight Vulpeculan aggression; humanitarian crisis; impact on world energy resources
PSYOPS	Conduct leaflet drops in BLA areas announcing CJTF commitment to Aurigan govt.
C2	Forward CP in Sagitta; TAC vicinity port of Vermillion
Force Protection	Priority of effort: APODs; SPODs; TAAs; LOCs
Decision Points	Decision point 2: CHOP MEB to LCC as operational reserve; commit operational reserve for employment vicinity TAI 1.
Branches	Move CJTF main body through Sagitta and attack Vulpecula to force withdrawal of forces from Perseus
CCIR	Priority: Decision criteria for decision point 2 (commitment of operational reserve)
Risk	High Risk – Mitigate risk through deployment of air and naval power

Figure 39. DST analysis to commit operational reserve

the operational reserve. This is such a critical decision because when the operational reserve is committed, the joint force risks culmination. If the reserve is committed too soon or too late, or along the wrong axis, the force may experience catastrophic failure. In the DST shown in figure 39, the JFC faces a multidirectional attack without the ability to employ a fully operational combined joint task force (CJTF). He risks losing his seaports of embarkation (SPOE), and a major enemy force is penetrating the forward edge of the battle area (FEBA). Does he commit the reserve or not? By framing the problem, the JFC knows intuitively to commit the reserve. However, even though this is a correct decision, how the decision is executed determines the decision's outcome. Framing the problem and determining the correct solution do not necessarily guarantee success. In this case, the war game has identified the early commitment of the operational reserve as a critical decision the JFC must make. The JFC and the staff examine this decision point and the information recorded on the DST from the war game and develop decision criteria that will be recorded in the DSM.

Figure 40 provides an example of a DSM format identifying the decision point, TAI, NAI, event, decision required, decision criteria, assets available for employment, and CJTF actions related to this example. This format deliberately separates decision criteria and CJTF actions into the information required for a warning order and the information required for an execute order. This builds flexibility for

Decision Support Matrix MCO

DP	TAI	NAI	Event	Decision Req	Decision Criteria	Assets	CJTF Actions
☆ FEBA A **WARNORD**	# 1	# 1	Vul forces engage FEBA A in strength along coastal AA and operational exploitation forces (OEF) moving toward coastal AA	Authorize JFLCC to issue warning order to operational reserve (MEB) to block penetration of FEBA A along coastal AA CHOP MEB to JFLCC	Vul forces threaten Bde-size penetration of FEBA A Vul OEF preparing to move 316th Armor Bde vicinity NAI 1 Perseus's forces at < 80% strength	LCC ACC MCC SOF Perseus and Auriga forces	Authorize preparation actions to CHOP MEB Issue warning order Revise CJTF guidance letter Submit ATO input CHOP MEB
Execute Order			Perseus's forces unable to block penetration of Vul forces FEBA A penetrated by Vul 1st Echelon Bde	Deploy MEB ashore Authorize JFLCC to commit MEB to block penetration of FEBA A along coastal AA	Vul Bde at > 70% strength has penetrated FEBA A Air and fires are insufficient alone to stop penetration Perseus's forces < 70% strength Perseus's reserve already committed	LCC ACC MCC SOF Perseus and Auriga forces	Issue execution order to CJTF and coalition forces Redesignate operational reserve

Notes: Assume maximum enemy movement rate (mounted) in contact to be .5 km/hr
Estimate Vul movement time between FEBA A and FEBA B to be 24 hrs
Estimate minimum preparation time for MEB to block Vul penetration of FEBA A to be 24 hrs

Figure 40. Decision support matrix with decision criteria for a warning order and execute order

the JFC and his components by allowing them to initiate preparations without committing forces prematurely.

It is impossible to forecast every decision a JFC must make; however, key preplanned decisions can be identified. Preplanned decisions are those the JFC and the staff know they must make. They include, but are not limited to, major branches and sequels discovered during the JOPP's COA analysis and war-gaming step, planning assumptions, priorities of effort, phase transitions, and time-sensitive targets.

We have already emphasized that each planning assumption requires a branch or sequel, but also each branch and sequel require a separate DST and DSM. If the planning assumptions are correctly identified and an assumption proves not to be true, the success of the entire operation or campaign is at risk.

The decision criteria embedded in the DSM provide a JFC an analytical tool for measuring risk and synchronizing the actions of friendly forces. They also assist a JFC in visualizing the timing of operations. If the DSM's decision is properly constructed, the JFC will have a much better gauge of when to issue a warning order and when to issue an execute order. The collection of DSMs formed from the decision analysis becomes a playbook of decisions that allows the JFC to review and study key decisions well before execution. Although JFCs will always have to rely on *coup d'oeil*, decision analysis aided by a DSM reduces the risks of ineffective or poorly thought-out plans and decision points.

The DSM also provides a structure for enhancing the effectiveness of CCIRs. JP 3-0 defines a CCIR as an information requirement identified by the commander as critical to facilitating timely decision making. A CCIR's two key elements are priority intelligence requirements (PIR) and friendly-force information requirements (FFIR). PIRs establish intelligence-support priorities that the commander and staff need to understand about an adversary or the operational environment. FFIRs identify information the commander and staff needs to understand concerning the status of the friendly force and supporting capabilities. The information derived from a DSM's decision criteria corresponds directly to the PIRs and FFIRs that the JFC needs to make a decision. Using the DSM decision criteria established in figure 40, figure 41 presents an example of how the DSM focuses CCIRs. The decision required is to commit the operational reserve to stop an enemy penetration of FEBA A along the coastal avenue of approach. The PIR and FFIR for this decision are taken directly from the DSM's

Decision Point	Decision Required	CCIR		Remarks
		PIR	**FFIR**	
★ 1	Commit operational reserve to stop penetration of FEBA A along coastal avenue of approach **WARNING ORDER:** Authorize JFLCC to issue warning order to op. res. (MEB) to block penetration of FEBA A along coastal AA CHOP MEB to JFLCC	Vul forces threaten Bde size penetration of FEBA A Vul OEF preparing to move 316th Armor Bde vicinity NAI 1	Perseus's forces at less than 80% strength	Latest time information of value: D+4
	EXECUTIVE ORDER: Deploy MEB ashore Authorize JFLCC to commit MEB to block penetration of FEBA A along coastal AA	Vul Bde at greater than 70% strength has penetrated FEBA A	Air and fires are insufficient alone to stop penetration Perseus's forces less than 70% strength Perseus's reserve already committed	

Figure 41. CCIRs' direct correlation to decision criteria established in the DSM

decision criteria and aligned with a time-sensitivity estimate indicating the latest time the information is of value. This focuses the collection and analysis of CCIRs against specific priority JFC decisions.

Throughout the JOPP the cognitive map is constantly refined to capture the JFC's operational vision. This map provides the JFC the ability to communicate his vision to political leaders, allies, combatant commanders, interagency organizations, intergovernmental organizations, nongovernmental organizations, and subordinate components. This communication mechanism links all three levels of war and serves as a blueprint for the operation or campaign. Figure 42 illustrates a refined cognitive map. The cognitive map and the analysis in the DST and DSM are essential elements for making effective decisions.

We must emphasize that not all decisions can be relegated to a nice, neat, preplanned format. Adversaries are complex, adaptive systems, and they are always capable of doing the unexpected. However, JFCs can shape emergent opportunities through both proactive and reactive operational maneuvers. The challenge is establishing a decision analysis framework that supports taking advantage of emergent opportunities.

CJTF Lines of Operation

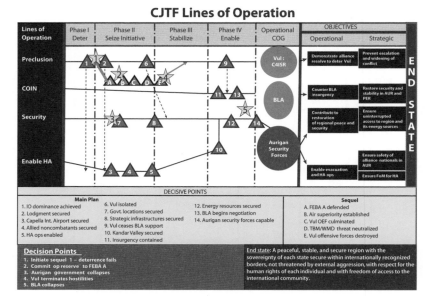

Figure 42. CJTF refined cognitive map

Shaping Emergent Opportunities

One of military history's most masterful illustrations of shaping emergent opportunities is Gen Ulysses S. Grant's decision analysis during the Civil War's Vicksburg campaign. Located 250 feet above the Mississippi River, the Confederate city of Vicksburg dominated transportation and supply along the river. It was a strategic choke point vital to the interests of both the North and the South and a crucial objective for the Union forces.

In the fall of 1862, Grant mounted an initial campaign to seize Vicksburg but failed after his logistics trains were destroyed at Holly Springs on 20 December, by Confederate general Earl Van Dorn. In spite of this setback, however, Grant immediately began analysis of three courses of action to seize Vicksburg. The first was to dredge a new channel in the Mississippi River allowing Union forces to bypass the artillery dominating approaches to Vicksburg. The second involved bypassing Vicksburg by opening an axis of advance through the bayous between Providence and the Red River in Louisiana. The third concentrated on finding an overland route on the west side of the Mississippi.[10] Although there was no formal doctrine at this time,

Grant's decision analysis demonstrated a mastery of operational design's principal concepts. He combined the constructs of center of gravity, lines of operation, and decisive points with deception to shape emergent opportunities.

At the beginning of Grant's second campaign to seize Vicksburg, the Union and Confederate forces were almost evenly matched. Both sides had approximately 50,000 men. Realizing this, Grant developed and executed a deception plan consisting of several diversions to confuse, fix, and realign Confederate force ratios. Grant ordered Maj Gen Frederick Steele to attack north of Vicksburg and destroy Confederate supplies and livestock. This attack, in conjunction with movements of Sherman's forces and Col Benjamin Grierson's famed cavalry raid, diverted Confederate attention north and concealed Grant's movements on the west side of the Mississippi. But Grant's boldest decision was to cut his own supply lines.

On 30 April 1863, Grant crossed the Mississippi River below Vicksburg at Bruinsburg (fig. 43) and was faced with two choices: halt and build up logistics or cut his supply lines and continue the attack. Grant seized the opportunity to continue the attack. His decision

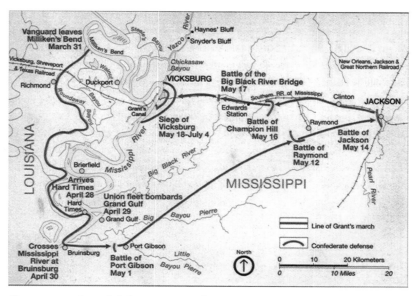

Figure 43. The Vicksburg Campaign. (Reprinted from Vicksburg National Military Park, map of the Vicksburg Campaign, http://www.nps .gov/vick/planyourvisit/park-maps-and-brochure.htm.)

analysis assumed that if he waited for supplies, his adversaries, Confederate lieutenant generals John C. Pemberton and Joseph Johnston, would be able to unite forces and reinforce Vicksburg with fresh troops.[11] Grant's methodical decision analysis process enabled him to envision how the campaign should unfold and established the conditions necessary to keep his adversaries off balance. His forces swept Confederate forces aside at Port Gibson on 1 May and continued on to Raymond where he again defeated the Confederates. At Raymond, Grant was confronted with Confederate forces to his west at Vicksburg and to his east at Jackson. At this critical point, he made the decision to halt his assault on Vicksburg and attack the city of Jackson to prevent the Confederate forces under Johnston from linking up with Pemberton's forces at Vicksburg. After defeating this threat to his east flank, Grant moved westward to seize Vicksburg. His forces defeated the Confederates at Champion Hill and again at the Big Black River Bridge, and by 17 May were ready to begin the assault on Vicksburg. Grant's initial assaults on Vicksburg on 17 and 22 May failed, and he was forced to conduct siege operations. The cumulative effect of the siege operations, however, took its toll on the Confederate forces, and on 4 July 1863, Vicksburg surrendered.

In spite of the delay caused by the siege, Grant achieved a tremendous victory at Vicksburg. For the South, the fall of Vicksburg was a monumental strategic and psychological loss. A large part of Grant's success was due to his decision analysis and his ability to formulate decision criteria. Grant used his preplanned decision analysis to develop viable responses to emergent opportunities.

Shaping emergent opportunities draws from the cognitive vision ingrained in operational design's decision analysis. The DST derived from the analysis and war gaming of COAs provides a critical framework for examining the effects of friendly and enemy actions on the campaign. Every action of a friendly or enemy force will most likely precipitate some form of reaction or counteraction. Anticipating these actions and counteractions is vital. JFCs and their staffs must be able to envision key decisions and establish clearly defined decision criteria. The creation of specific decision criteria not only provides a metric for commanders to assess decisions, but also focuses the collection efforts of the CCIR. This minimizes the passing of irrelevant information through the collection system and speeds up analysis. It is important to emphasize that no one can predict with 100 percent accuracy what an adversary will do. Adversaries are complex, adaptive, thinking systems. However, as JFCs and their staffs construct their campaign, a deliberate and disciplined analysis of potential decisions will provide an effective approach to exploiting emergent opportunities.

Notes

1. JP 1-02, *Department of Defense (DOD) Dictionary*, 370.
2. de Hartog, *Genghis Khan*, 86–98.
3. Greene, *33 Strategies of War*, 181.
4. US Army, *Army Strategic Planning Guidance*, 14–15.
5. Thayer, *War without Fronts*.
6. JP 1-02, *DOD Dictionary*, 98.
7. Ibid., 249.
8. Ibid., 361.
9. Ibid., 99.
10. Catton, *Grant Moves South*, 323; and Bearss, *Vicksburg Campaign*, 427.
11. Badeau, *Military History*, 162–63.

Chapter 5

Operational Design and Counterinsurgency

One of the common oversights embedded in doctrinal practices is applying foundational principles to all operational environments without assessing their relevance or impact. This is certainly the case in applying operational design to major conventional operations and counterinsurgency (COIN) operations. There are significant differences between these two types of operations. For both, as Clausewitz stated long ago, "the political object is the goal, war is the means of reaching it, and the means can never be considered in isolation from their purposes."[1] However, destruction of the enemy's armed forces in a COIN environment does not necessarily guarantee the attainment of the political objective.[2] This factor has a significant impact on a planner's approach to operational design. The planner must recognize the intricate nature of an insurgency's root causes and craft a design framework that supports the analysis of those complexities. Some traditional aspects of operational design, such as center of gravity and arrangement of operations by phase, may not relate or function in the same manner as they do in major conventional operations because of the COIN's complexity.

Joint Publication (JP) 3-24, *Counterinsurgency Operations*, defines insurgency as "the organized use of subversion and violence by a group or movement that seeks to overthrow or force change of a governing authority."[3] In contrast, COIN is the "comprehensive civilian and military efforts taken to defeat an insurgency and to address any core grievances."[4] The distinction between major conventional operations and COIN operations is not just the fight for governance. There is also a dramatic difference in the temporal dimension that affects operational design.

Most analysts acknowledge that a commitment to conduct COIN operations is often a commitment to decades of support. The root causes of insurgencies usually evolve over lengthy periods of time, and they are not going to be eradicated in a few months or even years. A recent RAND study of COIN campaigns since 1945 found that the successful ones lasted an average of 14 years and the unsuccessful ones an average of 11 years.[5] However, we must also understand that COIN is a race against time. The counterinsurgent must produce a balanced mixture of short-term and long-term results to solidify the support of the people. As David Galula states, "The counterinsurgent needs a convincing success as early as possible in order to demonstrate that he has the will, the means, and the ability to win."[6] As figure 44 indicates, over time the counterinsurgent's will to succeed and his level of ambition decline markedly. As time elapses and the cost goes up in lives and dollars, support for a COIN diminishes. This is especially true for democratic nations waging COIN operations. In any 10-year period, democratic nations undergo a series of business and election cycles that directly affect the level of COIN ambition and the will to win. Time will most often favor the insurgent. Consequently, a COIN design should shape the natural trajectory working against the will to win by showing immediate improvements in selected key areas. This requires a sophisticated design that fuses an understanding of the system with the ability to promote a sense of enduring commitment to the host nation. Maximum campaign effectiveness is achieved only when these two entities are integrated.

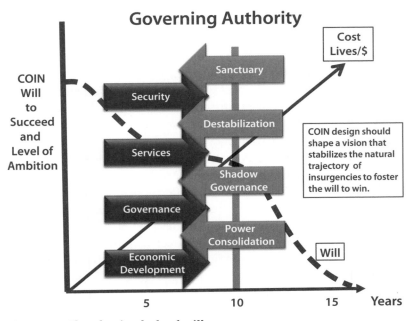

Figure 44. The classic clash of wills

The initial focal point to increase campaign effectiveness is to understand the system. Systems analysis is far from a new concept in the COIN environment. The United States used systems analysis extensively in the COIN effort during the Vietnam War. The Military Assistance Command Vietnam (MACV) defined a system as "a complex of interdependencies between parts, components and processes that involve discernible regularities of relationship." This definition also refers to a system as the interdependency between the complex and its surrounding environment.[7] Figure 45 provides a macro view of the systems analysis developed by the MACV.

This analysis was augmented with numerous research studies such as Douglas Pike's *War, Peace, and the Viet Cong* that provided a strategic perspective on the Vietnamese communist strategy of *dau tranh* (struggle). This strategy focused on the unremitting use of military and nonmilitary force over long periods of time in pursuit of an ob-

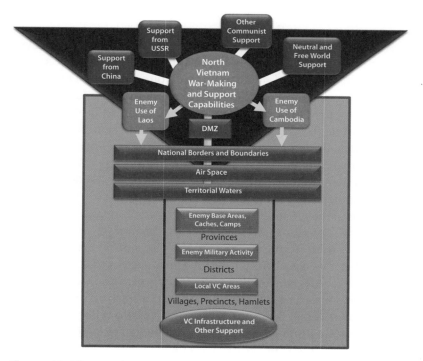

Figure 45. The MACV's systems analysis. (Reprinted from Combined Intelligence Center Vietnam, *The Enemy System: Subregion 1*, Research and Analysis Study ST 70-01, 24 August 1970, figure 1.)

jective and possessed a dual character: the armed struggle and the political struggle. Pike referred to the individual fighting the political struggle as

> the man in the black pajama of the Vietnamese peasant. His mission is to steal the people away from the government. His concern is almost exclusively control of the people, as distinguished from the big-unit war, where the concern is "control" of the enemy's army. He seeks to achieve this through programs designed to control the people, programs augmented and made possible by as much (and as little) military effort as is required.[8]

The intent of the MACV's analysis was to develop a conscious targeting effort against the total enemy system by acting on subsystem components. The enemy system was seen as complexes consisting of one or more base areas, travel routes, commo-liaison stations, and other elements that allowed it to functionally regenerate itself despite the loss of single elements. The systems model developed by the MACV encompassed the enemy subsystem, the Viet Cong Infrastructure (VCI), the military subsystem, the rear services subsystem, and an assessment of the system's strengths and weaknesses.[9]

Although the US failed in Vietnam, the MACV's systems analysis was an effective COIN tool. Between 1967 and 1972, the percentage of government-secured population rose from 42 to 80.[10] During that same time frame the Viet Cong's force strength declined by 50 percent, and by 1972 only 20 percent of the communist forces in South Vietnam were Viet Cong. The rest were North Vietnamese regulars or North Vietnamese in traditional Viet Cong units.[11] We should remember that South Vietnam's final capitulation was the result of a North Vietnamese conventional invasion backed by 165,000 soldiers, massive amounts of artillery, and over 600 tanks.

In spite of the US failure in Vietnam, systems analysis provides a crucial tool for today's COIN operations. Examining an insurgency begins with understanding "the system" (fig. 46). This requires defining what the system is and linking that system to campaign effectiveness. The framework for systems analysis is not just a political, military, economic, social, information, and infrastructure (PMESII) systems correlation. The systems analysis should reflect a comprehensive examination of the desired end state, related objectives, historical context, relevant actors, and effects of culture. Additionally, the framework should analyze potential barriers to COIN operations such as security, corruption, illiteracy, financial system, ethnic divisions, religious diversity, and enemy sanctuary. Identifying bar-

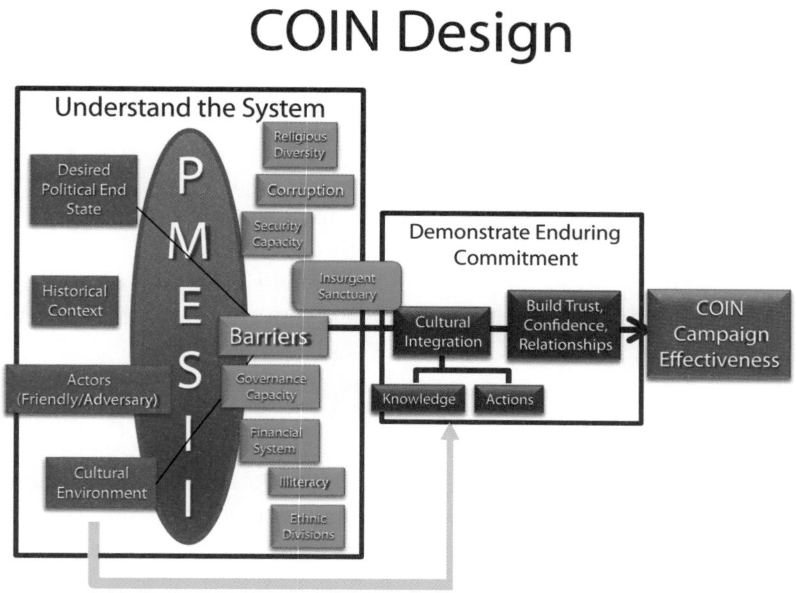

COIN Design

Figure 46. COIN effectiveness: understanding the system and demonstrating an enduring commitment

riers allows planners to scope the true context of the problem, analyze priorities, and determine a realistic balance of short- and long-term desired results.

Once COIN planners frame the system, the focus becomes how to engender a sense of enduring commitment to the host nation. COIN forces must earn the trust and confidence of the host nation. One method for accomplishing this is through cultural integration. The process of cultural integration provides a mechanism for guiding COIN forces in working within the host nation's culture. Planners assess the critical junctures where culture will have a dominant effect in galvanizing popular support for COIN efforts. Planners should use the historical, social, and anthropological information derived from the systems analysis along with human terrain mapping to develop a cultural integration process that links directly to policy actions.

There are three primary reasons why cultural integration is important. First, the insurgent uses the host nation's culture to communicate with the population and gain support. This use of culture empowers the clandestine infrastructure to project shadow governance. Second, when COIN forces operate outside of a culture, they alienate the population and render policy objectives ineffective. COIN forces must develop and disseminate narratives, symbols, and messages that resonate with the population's preexisting cultural system.[12] Third, governing authority is the focus for all COIN operations. The counterinsurgent reaches a position of strength only when his power is embedded in the host nation's political organization and it is firmly supported by the population.[13]

The challenge for COIN commanders and their staffs is how to digest the voluminous information emanating from the systems analysis and correlate that information to policies that provide the population with security, services, and governance. One method is recording the salient systems analysis information on an operational environment map. Using Afghanistan as an example, figure 47 depicts a generalization of historical, social, cultural, and security issues confronting the International Security Assistance Force (ISAF). The concept of operational environment mapping is neither a new nor a revolutionary idea. It simply provides a way to create a holistic vision for the commander by connecting systemic parts to the operational environment.

Although mapping the operational environment appears simplistic, it requires significant critical insight to distinguish the links between systemic parts. The Kajaki Dam in Helmand Province, Afghanistan, is an excellent heuristic example. This dam provides one of the most crucial services in Afghanistan—electricity. A 2006 survey conducted by the Asia Foundation found that 25 percent of Afghanistan's population listed the lack of electricity as their greatest problem. The only problem rated higher was unemployment. Progress in providing electricity has been systematically stalled. In 2001 approximately 6 percent of the population had access to electricity. By 2010 only 497,000 of the approximately 4.8 million households in Afghanistan had access to the national power grid.[14] This is well below the developmental goal of providing power to 65 percent of urban and 25 percent of rural households.

Figure 48 presents a hypothetical map of the Kajaki Dam's operational environment. The map highlights key PMESII factors and the tremendous problems associated with security. But the map fails to show the seams between the interagency vision, the host-nation government's requirement for immediate progress, and the military's ability to provide security. The United States Agency for International Development's (USAID) vision for the dam is a long-term objective to provide cheap electricity to areas of Kandahar, Kajaki, Sangin, Musa Qala, and Lashkar Gah. The Afghan government, however, is under intense pressure to supply electricity to its constituents now, and the failure to provide this service undermines the government's authority to govern.

The holdup is the military's ability to create a secure environment, not only in the area around Kajaki, but also in the surrounding districts where the hydroelectric plant's power cables traverse. Since 2006 numerous military operations have supported the Kajaki Dam development. Progress, however, was hindered because a third turbine required to generate additional power could not be moved along the route to the dam. In 2008 over 2,000 British soldiers escorted the third turbine generator to the dam site. Despite this, the turbine was still not operational in 2010 because the cement and other materials required to emplace the turbine could not be transported along the route to Kajaki. The other potential answer to this problem is to supply the population in these areas with generator banks. This form of energy is more expensive, but it is easier to construct, requires less security, and is more responsive to the people's needs.

Ironically, the beneficiary of the Kajaki Dam dilemma is the Taliban. The failure to make the dam operational highlights the Government of the Islamic Republic of Afghanistan's (GIROA) inability to provide the basic essentials of security and services to its people. Additionally, the Taliban's control of the surrounding districts indirectly presents it with a dual source of income to fund its operations. It diverts energy from the Kajaki hydroelectric plant to irrigate poppy fields for opium production and charges the local citizenry for the electricity that does come from the plant. The Helmand provincial government estimates it loses $4 million a year in electricity revenue to the Taliban.[15]

The Kajaki Dam illustrates the need to connect systemic parts, especially between the different levels of war. Creating a map of the operational environment empowers cognitive vision, but it does not present a means to analyze interconnecting systemic parts. One solution to this is combining the map of the operational environment with traditional aspects of operational design.

Op Design: Systems Analysis

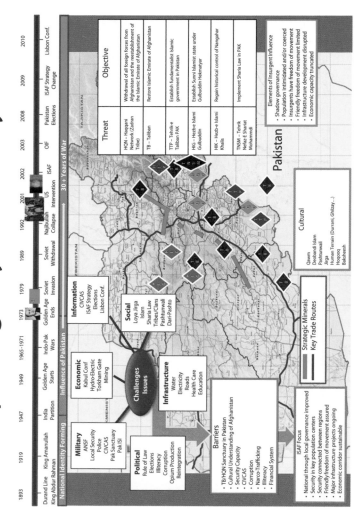

Figure 47. Map of the operational environment

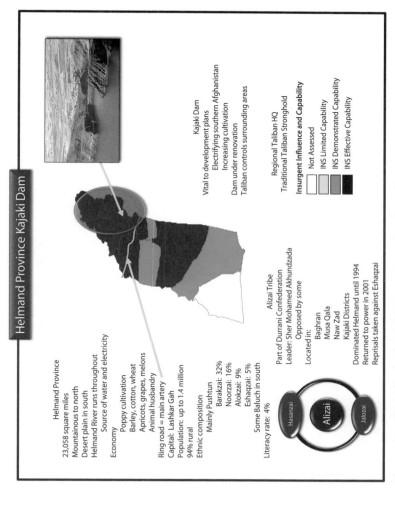

Figure 48. Hypothetical Kajaki Dam operational environment map

Relationship between the Traditional Aspects of Operational Design and Irregular Warfare's COIN Operations

Assessing any type of operation begins with an understanding of the levels of war or operations. Many planners talk past each other because of confusion over what level of war they are discussing, and nowhere is it more confusing than in COIN operations. No two insurgencies are the same. They differ in their root causes, operational environments, and insurgent forces. Additionally, insurgent actions at the tactical level draw immediate attention from all levels of war, and this phenomenon blurs responsibilities, interagency integration, and command relationships. Figure 49 presents a traditional view of the levels of war and corresponding operational designs. COIN operations have an inherent complexity, and insurgencies cannot usually be resolved through quick, decisive military action. As a result, there are different types of interactions among the levels of war. These inter-

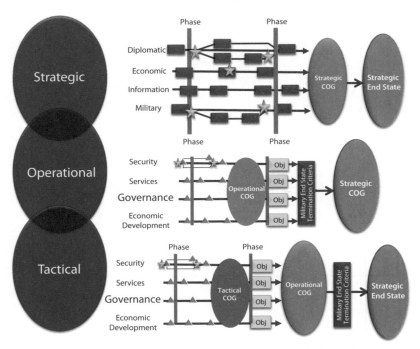

Figure 49. Traditional operational design and the relationships among levels of war

actions require much more detailed coordination than the traditional model and change the way we visualize operational design elements.

In COIN operations, the interrelationships among the levels of war move away from a hierarchical structure to a much more integrated and complex arrangement similar to figure 50. The reason for this shift is the pervasive complexity embedded in a COIN's operational environment and the scope of the interagency interactions necessary for success. In this model, the tactical level of war rests on top of the operational and strategic levels because tactical actions are what everyone sees first, and those actions can have immediate effects on both of the other levels. For example, on 30 September 2010 two NATO helicopters crossed into Pakistan's Kurram tribal region and fired on a Mandata Kandaho border patrol post, killing three Pakistani soldiers. Within hours, Pakistani authorities ordered a halt to all trucks and oil tankers ferrying supplies to the ISAF's forces through Afghanistan's Torkham Gate checkpoint.[16] This tactical confrontation resulted in a 10-day blockade of the Torkham Gate area and the destruction of an estimated 150 trucks by Pakistani and Taliban attackers.[17] This situation was serious enough to require international mediation at the strategic level to reach resolution.

The integrated arrangement shown in figure 50 also supports the most common view of the center of gravity in COIN environments. In major conventional operations, identifying COGs at each level of war provides a mechanism for focusing military forces during the campaign's phases. In COIN operations, the complexity of coordinating military, interagency, intergovernmental, and nongovernmental organizations may necessitate the identification of a single COG to achieve a focus of efforts. Military planners naturally tend to concentrate their COG analysis on the enemy's forces. The problem for planners in many COIN contexts is that there may not be a single entity that constitutes a homogenous insurgency. In Iraq, military forces have been confronted by Sunni, Shia, and al-Qaeda operatives and numerous criminal networks. The same is true for Afghanistan, where insurgents come from multiple groups, including the Quetta Shura Taliban, the Haqqani network, the Hezb-e-Islami Khalis (HIK), Hizb-l Islami Gulbuddin (HIG), and the Tehrik-i-Taliban (TTP). Each of these groups has different objectives. Although it appears counterintuitive, identifying a single COG to support COIN operations may actually provide better focus.

One of the most often cited COGs in COIN operations is the host nation's population. This concept directs the focus of COIN opera-

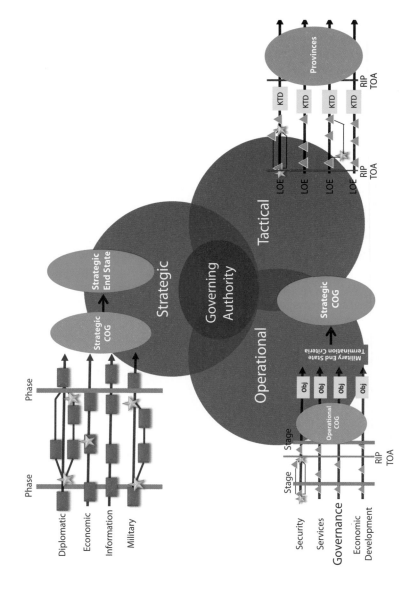

Figure 50. COIN design and the relationships among levels of war

tions at security, services, governance, and economic development for the people. What this concept misses, however, is the fact that COIN forces can focus almost exclusively on the population and fail to subdue the insurrection. COIN forces can provide security, services, and even economic development for the population and not succeed because the people have too little confidence in the host nation's governing authority. Any gains in security, services, and economic development are ephemeral unless they inspire confidence in the governing authority. Additionally, as figure 50 depicts, governing authority must be ensured simultaneously at all three levels of war to produce success. The strategic level of war has the responsibility for transforming or developing a form of government that is legitimate in the eyes of the people and coordinating the international support for that government. The operational level coordinates security, stability, and economic development operations throughout the nation to highlight the host nation's ability to govern effectively. Last, but no less important, the tactical level has the burden of execution. The face at the forefront for all three levels of war should be the host nation.

Focusing on governing authority also functions similarly to a Janus strategy. On one side, COIN forces support and facilitate the host nation's governance. But on the other side, COIN operations should also concentrate on neutralizing the insurgent's governing authority. Using this perspective, the strategic level focuses on denying international support and sanctuary to the insurgents. The operational level coordinates and directs military operations aimed at neutralizing the insurgent's clandestine infrastructure and denying the ability to establish shadow governance. The tactical level once again has responsibility for executing the operations.

The COG is not the only area in which operational design is applied differently for COIN operations than for major conventional operations. Because of the temporal dimension of war, the arranging of operations is another area in which there are critical differences. Although major conventional operations can last years, COIN operations may last decades. As a result, arranging operations into phases may not be effective. If a COIN campaign's stability-operations phase lasts years, how does the operational level of war coordinate and direct operations against an adaptive adversary with friendly forces and headquarters rotating in and out of a theater? In that type of environment, the use of phases makes it almost impossible to maintain continuity of effort, develop assessments, and simultaneously adjust to an ever-changing operational context.

The answer is to break phases into stages and follow the relief in place/transfer of authority (RIP/TOA) process (fig. 51). Stages allow phases to be broken into smaller, more manageable portions of time to direct and adjust operations. One of the greatest obstacles to the ongoing efforts in Iraq and Afghanistan is the lack of continuity between units rotating in and out of theater. Whether it is an Army brigade combat team (BCT) replacing a Marine regimental combat team (RCT) in al-Anbar, Iraq, or an Army BCT replacing another Army BCT in Afghanistan's Regional Command (RC) East, the operational-level headquarters must subdivide phases to ensure transitional continuity. Units have been rotating in and out of Afghanistan and Iraq since 2002, and as the commanders have changed, the vision for success has also changed.

In 2008 a BCT from the 101st Air Assault Division rotated into Afghanistan and began work on the Khost-Gardez road system. The objective was similar to the Kunar model explained in David Kilcullen's *Accidental Guerrilla*.[18] The purpose of the road was not simply to develop infrastructure. The road was a means to project combat power for security, stimulate economic growth by creating access to markets, and simultaneously demonstrate governing authority. In 2009 the BCT's RIP/TOA occurred; the next BCT rotated in and focused on dif-

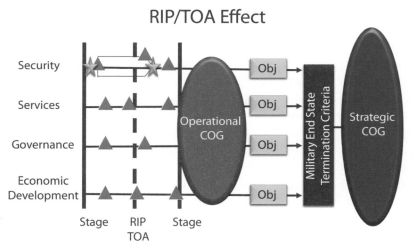

Figure 51. RIP/TOA has significant impact on objectives and operational continuity

ferent priorities. Work on the road system came to a virtual standstill. When the BCT from the 101st Air Assault Division rotated back to Afghanistan in 2010 and found that their progress on the road had stopped, they immediately refocused on the road. The purpose of this anecdote is not to judge the commanders from either of the BCTs, but to highlight the challenges of maintaining continuity during a lengthy phase. Unfortunately, this example is far from unique, and a breakdown in continuity occurs not only between military units but also between agencies and multinational organizations. The danger is that this inconsistency degrades the host nation's confidence in the United States' ability to foster an enduring international commitment.

Establishing continuity and consistency in a COIN operation depends on three salient factors: a standardized structure for operational design at all three levels of war, an in-depth appreciation for local security, and the ability to understand and conduct assessment. The notion of developing a standardized structure for operational design faces significant opposition because many fear it will lead to prescription and a checklist mentality. There is a marked difference, however, between dictating how a commander develops a design and standardizing how the design is recorded and used for synchronization across the levels of war. It is unhealthy to put constraints on how a commander develops a design. A commander must have flexibility in formulating a design so that it supports his or her vision for execution. But not establishing a framework for operational design is similar to conducting the JOPP and not recording the end product in accordance with the five-paragraph format for plans and orders. Furthermore, an operational design must be understood not only by US and multinational military forces, but also by the interagency and multinational organizations participating in the COIN. It should facilitate synchronization and leverage the national instruments of power at all three levels of war. This cannot be achieved without a common framework that serves as an integrating mechanism.

COIN design at the strategic and operational levels is closely related to the design developed for major conventional operations. The core of the design is based on the relationships among the end state, objectives, effects, and COG(s) and the identification of decisive points and lines of operation.

Figure 52 depicts a notional structure attempting to align actions at the strategic and operational levels of war based on a common framework. This cognitive map is a starting point to ensure both levels

COIN Strategic Decision Analysis

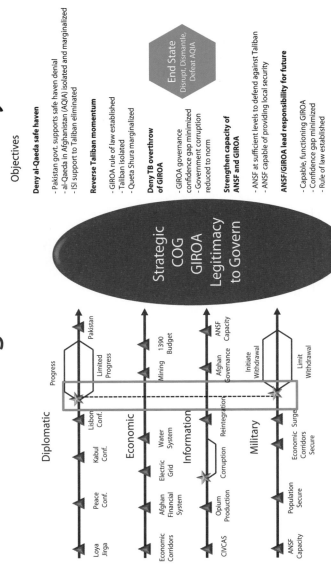

Figure 52. Synchronizing operational design relationships among levels of war

COIN Operational Decision Analysis

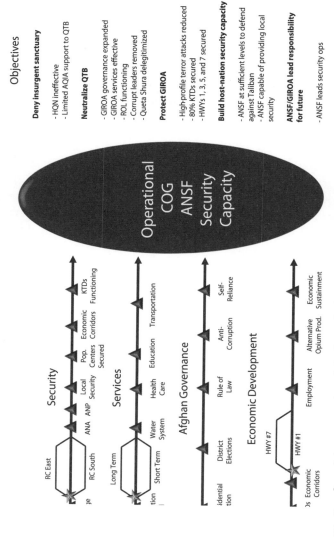

Objectives

Deny insurgent sanctuary
- HQN ineffective
- Limited AQIA support to QTB

Neutralize QTB
- GIROA governance expanded
- GIROA services effective
- ROL functioning
- Corrupt leaders removed
- Queta Shura delegitimized

Protect GIROA
- High-profile terror attacks reduced
- 80% KTDs secured
- HWYs 1, 3, 5, and 7 secured

Build host-nation security capacity
- ANSF at sufficient levels to defend against Taliban
- ANSF capable of providing local security

ANSF/GIROA lead responsibility for future
- ANSF leads security ops

Operational
COG
ANSF
Security
Capacity

Security
RC East
RC South
ANA ANP Local Security Pop. Centers Secured Economic Corridors KTDs Functioning

Services
Long Term
Short Term
Water System Health Care Education Transportation

Afghan Governance
District Elections Rule of Law Anti-Corruption Self-Reliance

Economic Development
HWY #7
HWY #1
Economic Corridors Employment Alternative Opium Prod. Economic Sustainment

Figure 52. Synchronizing operational design relationships among levels of war (continued)

of war are working in conjunction with one another and not working at cross purposes. It identifies key decision points, relationships among lines of operation, points at which the two levels of war are not synchronized, and the potential impact of improperly aligned stages and RIP/TOAs.

At the tactical level of war, complexity has an enormous impact on the development of the cognitive map. Tactical commanders face a great diversity of challenges, such as different insurgent groups operating in their area of operations, ethnic and tribal enclaves, and vast differences in area resources and host-nation political leaders' capabilities. To meet the challenges emerging out of the US COIN experiences in Iraq and Afghanistan, the organizational structure of the combined joint operations area (CJOA) is evolving. This is clearly seen in Afghanistan, where the CJOA structure is based on regional commands (RC), key terrain districts (KTD), and area-of-interest districts. This type of CJOA organization is not a new concept. It has been used in varying degrees since ancient times. This organizational approach, however, provides a fresh perspective on battlespace ownership and focusing operations. In this concept, the operational-level headquarters oversees a series of regional commands. The regional commands coordinate all regional civil-military activities conducted by the military elements in a number of different provinces. An illustration of this is seen in the ISAF's regional command organization of Afghanistan (fig. 53).

To maximize the effectiveness of COIN operations, each of these regional commands has undergone a district-level key terrain assessment to identify key terrain districts and area-of-interest districts. KTDs contain concentrated populations, major centers of economic productivity, key infrastructure, and key commerce routes. Area-of-interest districts exert influence on KTDs and facilitate information collection and operational resource support.[19] Additionally, the ISAF has linked the KTD system to economic corridors that support the vision for economic development. Figure 54 illustrates Afghanistan's KTDs, area-of-interest districts, the Ring Road, and the border crossing points that generate the majority of Afghanistan's revenue.

The ISAF and its operational-level headquarters, the ISAF Joint Command, have a specific set of selection criteria for designating KTDs and area-of-interest districts. However, planners should understand that the physical location of KTDs may also be an important consideration for selection. In *Counterinsurgency Warfare: Theory and Practice*, David Galula describes the optimum geographic environments

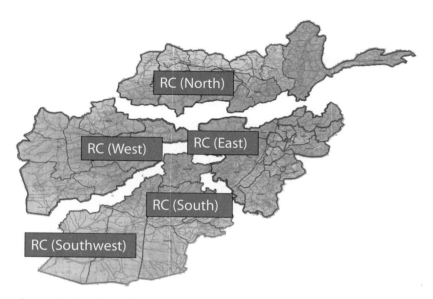

Figure 53. ISAF's regional commands

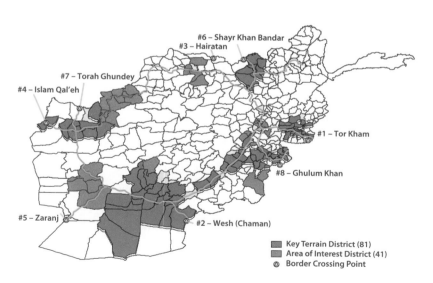

Figure 54. Key terrain districts in Afghanistan

and physical shapes for both insurgencies and COINs. He states that the most favorable shape for an insurgency is a land-locked blunt star and for COIN operations an island shaped like a pointed star.[20] Although Galula does not elaborate on the rationale for these shapes, the primary consideration is undoubtedly security of the population. What is significant about Galula's main point is that KTDs should be positioned where they collectively provide maximum security for the population. A commander may not have the luxury of conducting COIN operations on an island shaped like a pointed star, but he or she may be able to artificially create those conditions.

Doctrinal changes are also affecting the articulation of tactical-level cognitive maps. Recently JP 5-0, *Joint Operation Planning* (revision final coordination), introduced the term *lines of effort* (LOE) to the joint lexicon. LOEs establish operational and strategic conditions by linking multiple tasks and missions.[21] If we combine the impact of RIP/TOAs on operational continuity, KTDs, economic corridors, and LOEs with the overarching focus on governing authority, the cognitive map at the high tactical level may resemble figure 55. In this model, the operational-level headquarters issues guidance to RCs on objectives using stages and RIP/TOAs to delineate overlapping responsibilities between units. The stage defines what the common objectives/tasks are and the RIP/TOA describes the specific KTD conditions units must achieve to meet their responsibilities towards those objectives/tasks. The tactical units assuming responsibility for the RC then use

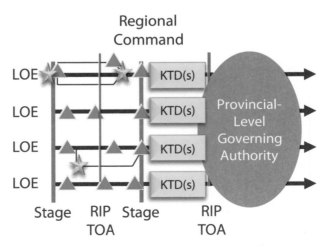

Figure 55. Tactical-level COIN design

that guidance to develop LOEs aimed at the directed KTD tasks. The cumulative impact of the LOEs and KTD tasks should be directed at enhancing the provincial-level governing authority.

Some examples of tactical-level LOEs are:

- Expanding governance of KTDs along Highway 7.
- Securing freedom of movement along Highway 1 to facilitate economic opportunity for KTDs A, B, and C.
- Securing the economic corridor from Gardez-Khost (GK) to Highway 1 to increase stability for KTDs D, E, F, and G.
- Building local security force capacity in KTDs H, I, and J.

The overall objectives of providing a common structural framework for operational design are continuity and unity of effort. Units and agencies responsible for executing COIN operations need as much clarity as possible. This can happen only with an understanding of everyone's roles and a vision for integration. Additionally, a common framework affords the ability to visualize risk and leverage interdependencies between agencies to mitigate that risk.

Centrifugal Force of Local Security in COIN Operations

The chances of two insurgencies being the same are nil, but local security is a decisive point that deserves planners' explicit attention. Security is the first principle of warfare, and nowhere is it more important than in COIN operations. Almost every aspect of COIN depends either directly or indirectly around local security. As a result, local security is a crucial decisive point along the security line of operations. In spite of this fact, local security remains an area plagued by mistakes. When COIN forces build host-nation security capacity, there is a consistent tendency to concentrate almost solely on conventional forces. Local security in actuality requires a deliberate layering of forces. Conventional forces are good at conducting strike operations in insurgent-controlled areas and assisting in the establishment of security in contested areas. But conventional forces are not optimal for static security or law and order.

The traditional defense pattern for successful COIN operations (fig. 56) uses conventional forces to establish a security belt that enables

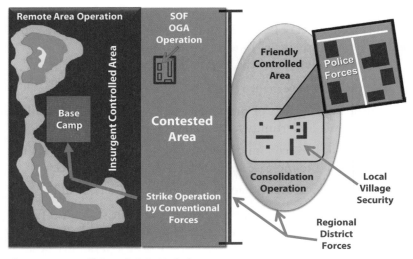

Figure 56. Traditional COIN defense patterns

local security to take place. This is done in concert with special operations forces that prepare insurgent-contested areas for the introduction
of local security forces. After this foundation is established, local security forces assume security responsibility for consolidation operations
and the introduction of police forces. Local security forces tie into police operations and bring the population under government control.
This widespread dispersion of forces provides better protection for the
people and frees conventional forces to strike insurgent refuges.

The classic mistake is handing off responsibility for local security
to police forces before the necessary conditions have been established
for their success. Handing off local security to an unprepared police
force is a critical and unforgiving vulnerability. A key illustration of
this is the Brookings Institution's data on Afghan security force casualties, 2007–09 (table 3).[22] In addition to those casualties, Afghanistan's
Ministry of the Interior reported 595 police officers killed and 1,345
wounded during the first six months of 2010. There are numerous
reasons for the disparity of casualties between the Afghan National
Army (ANA) and the Afghan National Police (ANP). Two of the primary causes are training and equipment. The ANP are neither trained
nor equipped to face the security environment they are being thrust
into. The ISAF has made Herculean strides to correct these deficien-
cies, but there is still a missing dimension in Afghanistan's security—
the intermediate layer of local security forces.

Table 3. Annual security-force fatality totals

	2007	2008	2009
Afghan National Army	209	226	282
Afghan National Police	803	886	646

Known by names such as *territorial defense* and *civil defense*, local security has a significant effect on all three levels of war. At the tactical level, once conventional forces have established an initial security belt, local security enables stabilization to take place. It does this by insulating the population from insurgent demands for food, money, medicine, and other resources and by reducing the fear of insurgent reprisals for passing intelligence information to COIN forces. This not only increases insurgent vulnerabilities, but also generates confidence in the reconstitution of governing authority.

At the operational level of war, local security provides the foundation for expanding infrastructure and lines of communication. There is an unmistakable relationship among infrastructure, lines of communication, and a government's ability to project power. The significance of this relationship is seen in the priority and frequency of insurgent attacks on infrastructure and lines of communication. Underdeveloped lines of communication favor insurgencies, and the sabotage of critical infrastructure constitutes a fundamental element in most insurgent strategies. In most cases, insurgents do not have the military capability to overtly gain power. They accomplish this indirectly by destroying the government's economic base. This accentuates economic crises, creates unemployment, inflates the cost of living, and perpetuates a fertile base for recruitment.

During the insurgency in El Salvador (1979–92), one of the Farabundo Martí National Liberation Front's (FMLN) key objectives was preventing productive economic activity. A captured insurgent document outlined the following areas for systematic destruction: fuel pipelines, electricity grids, railroads, telephonic communications, and agricultural export crops.[23] The FMLN's targeting of infrastructure had another deliberate and insidious effect. It generated capital flight and inhibited international investment. From 1979 to 1980, over $2 billion fled El Salvador for Miami.[24]

The ability to limit and prevent the types of insurgent depredations described above rests with the realization of the intrinsic impact that local security forces have on the operational environment. Investing in local security is one of the wisest investments COIN forces can make. However, local security's contributions are not confined to the dimension of physical security at the tactical and operational levels of war. Local security can also serve as a strategic COIN function. If implemented properly, local security is a catalyst for inducing the populace to support the host-nation governing authority. Individuals who commit themselves to local security, especially at the village level, demonstrate a clear rejection of the insurgent movement.[25] This commitment decisively strips away support for the insurgency and imperils the insurgent's strategic center of gravity, the ability to exert governing authority.

Models for Implementing Local Security

There are numerous models for establishing intermediate local security forces. However, the two basic ones are arming local tribesmen, similar to the Sons of Iraq (SOI), and the Vietnam-era regional/popular forces model. Both have advantages and disadvantages, and both function best in specific operational contexts. The SOI are Iraqi civilians who have voluntarily allied with US forces to provide security against insurgents and militias at the local level.[26] The movement began slowly during 2005 in al-Anbar Province. Known as the al-Anbar Awakening Council, groups of Sunnis, some of them former militants, banded with US military forces against al-Qaeda. The US forces gradually turned local security responsibility over to these groups, paying members $10 per day. By late 2007 the SOI included both Sunnis and Shiites, and groups were active in eight of Iraq's provinces, with a membership numbering over 73,000.[27] The SOI had over 100,000 by 2008. Table 4 shows the status of the SOI in 2009.

The impact of the SOI combined with the military surge is readily evident. Civilian casualties dropped from a high of 34,500 in 2006 to 3,000 in 2009, and US military fatalities dropped from 905 to 149 during the same time frame.[28] The challenge with the SOI model is its ad hoc nature. It worked in a specific environmental context; however, its long-term effect remains unknown. The Iraqi central government is reticent to integrate these groups into the national-security force structure and is gradually phasing them out of their security role.

Tens of thousands of SOI members are still without full employment, despite the Iraqi government's pledge to provide police posi-

Table 4. Status of the SOI as of September 2009

Location of SOI Group(s)	Date of Transfer	Number of SOIs
Baghdad and immediate vicinity	1 October 2008	~51,000
Diyala/Qadisiyah Provinces	1 January 2009	~11,000
Anbar/Babil/Wasit Provinces	1 February 2009	~12,000
Ninewa/Tamim/Salahuddin Provinces	1 April 2009	~20,000

Reprinted from Michael E. O'Hanlon and Ian Livingston, *Iraq Index: Tracking Variables of Reconstruction & Security in Post-Saddam Iraq* (Washington, DC: Brookings, 31 October 2010), 9.

tions and vocational training. In March 2010 only 40 percent of SOIs were fully transitioned into full-time jobs.[29] Distrust between the SOIs and the Iraqi central government has been building consistently with SOI complaints of late paychecks, failure to pay salaries, and arrests of its members.[30] Many fear that these conditions are generating an incensed Sunni population that is potentially dangerous to the future stability of Iraq. Evidence is seen in numerous allegations of SOI members defecting back to al-Qaeda. Arming tribes can be an expedient approach to fostering local security; however, it can be high risk and in the long run work against stability. The key is ensuring that the host nation keeps its pledges.

The Vietnam-era regional/popular forces model is a much more deliberate and integrated method of fostering local security. The regional forces (RF) were company-sized elements that operated in a district or province. The popular forces (PF) operated as platoon-sized forces and were usually assigned to a specific village or static security task such as guarding key infrastructure, roadways, and waterways. Both of these forces worked in conjunction with the national police. Although they remained predominantly in their home provinces and villages, the RFs' and PFs' principal loyalty was to the central government.

The basic concept of using these forces predates the American involvement in Vietnam, but it was not until after 1965 that their greatest expansion occurred. In 1965 the total force structure for RFs, PFs, and national police was 320,000. By 1972 these forces numbered 664,000.[31] Their work was exceptionally dangerous, and they suffered horrendous casualties. However, they were also very effective. Evidence to support their effectiveness is seen in figure 57 showing the forces and their corresponding responsibility for the security of the

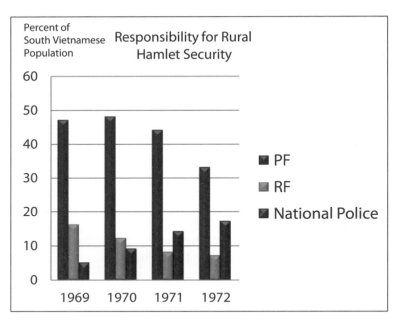

Figure 57. RF/PF responsibility for rural hamlet security. (Based on data from Hamlet Evaluation System Computer Tapes, 1969–72, in *Southeast Asia Statistical Summary,* Office of the Assistant Secretary of Defense (Comptroller), 14 February 1973.)

population. At first glance it appears that the RF/PF effectiveness diminishes between 1969 and 1972, but closer analysis shows the opposite effect on local security. As the RFs' and PFs' responsibility for security declined, they were being replaced by greater numbers of national police, indicating a more secure environment. In COIN operations, security forces transferring their responsibility for securing the population to police forces is a major sign of a return to normalcy. Between 1969 and 1972, the national police's responsibility for the security of the population increased more than three times from 5 to 17 percent. This relationship also correlates with the South Vietnamese government's expansion of security.

Another indicator of the RFs' and PFs' viability is their impact on the Viet Cong Infrastructure, the clandestine organization that commanded the majority of communist operations, established shadow governance, and conducted most of the terrorism against the population. Together the RFs and PFs accounted for a larger share of VCI killed or captured than any other force. In 1970 they were responsible for 50 percent of the VCI killed or captured and in 1971 approximately 39 percent. As Thomas Thayer points out in *War without Fronts*, the RFs and PFs were the most cost-effective forces of the Vietnam War. They secured more territory than any other force, increased friendly-force ratios, and inflicted between 23 and 33 percent of the total communist combat deaths. They accomplished this while receiving only 4 percent of the total war budget.[32]

Effectiveness in the RF/PF model does not originate from simply emplacing another layer of security at the local level. Planners need to understand efficiency and how to generate the local security forces' commitment to the host nation's governing authority. An option that is often ignored is establishing a mechanism for recycling conventional-force manpower. Under this concept conventional force members that have completed their term of service are either placed into or recruited for local security. Many nations operate with militaries founded on a system of conscription that obligates individuals to a certain number of years of service. Once they complete their term of service, these individuals return to their homes without further obligation. The cost of not leveraging this efficiency is significant. Between mid 1981 and January 1983—during El Salvador's Civil War— over 7,000 Salvadoran soldiers were trained at Fort Benning, Georgia. By June 1983 only half of these soldiers were still on active duty, and of the ones trained in 1981, only 15 percent remained.[33] The failure to harness this manpower for local security was amplified by the cost of the soldiers' training. The cost of training El Salvador's Bellosso battalion, 500–600 men, was $8 million.[34]

Galvanizing the population to support local security is an exceptionally difficult task. Local security patrols represent a direct challenge to insurgents for control of the population, making them high visibility targets. The simplest and perhaps most effective method of gaining popular support is by expanding the populace's material reliance on the government. In many situations this can be done at a negligible cost. Areas worthy of investment include:

- medical care for wounded personnel,
- immunizations, aspirins, and other common drugs,
- agricultural information on crops, agronomy, and animal husbandry,
- education for children and adult literacy programs,

- well digging, and

- loans for fertilizer and water pumps.

The RF/PF model possesses a number of distinct advantages; however, it also has a number of disadvantages. The creation of RF/PF local security forces is a slow and resource-intensive process. It requires additional manpower, trainers, and equipment. Its principal strength is that it can be shaped to demonstrate loyalty to the central government, but that loyalty must not be allowed to devolve into assassination squads. During El Salvador's civil war, the civil defense force, Organización Democrática Nacionalista (ORDEN), created many of the conditions that ignited the war and prolonged it. During 1980 alone, ORDEN was suspected of committing 861 political murders.[35]

Structuring the host-nation security capacity is one of the most significant decisive points in a COIN operation. Whatever model planners use, they must understand the gravity of local security and match it to the operational context. Matching the wrong model to the wrong context can have catastrophic consequences.

Making Decisions and Supporting Decision Analysis in the COIN Environment

Decision making and decision analysis are different in COIN environments than in major conventional operations in several important ways. The first is temporal. During major conventional operations, most command-level decisions evolve in a time-compressed planning-decision-execution (PDE) cycle. Moving large forces and changing component orders such as the air tasking order often require decisions to be adequately executed within a 72- to 96-hour period. In a COIN environment, most decisions emanate out of an extended PDE cycle. The principal reason for this is that very few military actions in a COIN mandate immediate responses. Most decisions in COIN operations are long-term projections involving the implementation of governance policies, the oversight of economic development projects, and the improvement of basic services. There are exceptions, however, including the commitment of a quick-reaction force, the kinetic engagement of a high-value target, and the daily battle for dominance in information operations.

Decision making in COIN operations also differs in the coordination, approval authority, and time required to obtain approval. In a COIN, effectiveness stems from the comprehensive involvement of the host nation, allies, intergovernmental organizations, other US government agencies, and, in some cases, key nongovernmental organizations. Many decisions are made by consensus, and approval authority can necessitate the involvement of national and multinational leaders.

Another difference is that targeting decisions may be more complex in a COIN environment than in conventional operations. An illustration of this is the kinetic engagement of a high-value insurgent target. In today's environment, the process is not as simple as identifying, nominating, locating, and engaging the target. Attacking a target entails compiling evidence against the target, reviewing the legality of engaging the target, and conducting a thorough collateral damage estimate. Additionally, the decision maker should analyze what removal of that target means to the insurgent system. Timing is everything, and in some cases removal of a high-value target can actually work against COIN efforts—for example, if the high-value target is making overtures toward reintegration. Of course, targeting is not confined to kinetics. Another kind of targeting is the removal of corrupt officials. As for kinetic targeting, the decision maker must gather legal evidence and analyze the effect of removing an official on the host nation's political system. Counterinsurgents must also determine whether to bring that official to trial, relieve him or her of duties, or marginalize his or her influence by empowering other officials.

COIN Decision-Support Tools

The diversity of decisions that confront COIN commanders demands the development and use of a variety of skillfully crafted decision-support tools. One of the most critical is the decision-support template, just as it is in major conventional operations. A COIN DST is very similar to the ones used in major conventional operations. It consists of an operational diagram depicting the major operations and objectives of the CJOA and a comprehensive synchronization matrix (fig. 58). In a COIN environment, synchronization may require the identification of such details as Gregorian and Persian years,

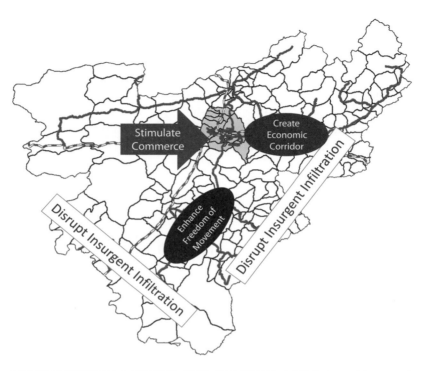

Gregorian Year	2010												2011											
Month	1	2	3	4	5	6	7	8	9	10	11	12	1	2	3	4	5	6	7	8	9	10	11	12
Persian Year	1388			1389												1390								
Season	Winter		Spring		Summer			Fall					Winter			Spring			Summer			Fall		
Key Events	1389 Budget				Kabul Conf.			Parliamentary Elections			LISBON Conf.		1390 Budget			Reintegration								
Key Dates	Al-Hajj Ashura			Nauroz		Victory of Muslim Nation (28 APR)		Ramadan (11 AUG – 9 SEP) Independence Day (19 AUG)					Al-Hajj Ashura			Nauroz			Victory of Muslim Nation (28 APR)					
Haqanni HIG HIK TTP			**Quetta Shura**																**Reconciliation Meeting**					
ANSF						**ANP: 109K by Oct '10**										**ANA: 171.6K by Oct '11**								
IJC	OMID I												OMID II											
RC South		Mastarak																						
RC Southwest					Hydro Power																			
RC East			KG Road								GG Road				Revenue Growth Torkham Gate									
RC Center			Security Kabul Conf.																					
RC North			KTDs 1-5																					
RC West							KTD Economic Corridor																	
Rip/TOA	RC South						RC Southwest												RC East					

Figure 58. Notional COIN DST

seasons, budgets, key cultural dates, major insurgent and friendly-forces activities, and the roles of other governmental agencies.

The DST assists commanders and staffs with identifying key interactive links and critical junctures for assessment. This is vitally important because the lengthy nature of COIN operations makes it exceptionally difficult to pinpoint key decision points and adapt operations to an ever-changing operational environment and adaptive adversaries.

Although we can war-game certain aspects of COIN operations, real analysis must be focused on the collection and examination of longitudinal data. This data should focus primarily on the insurgent's effectiveness in exerting governing authority and the population's attitude toward the host nation's governing authority. Longitudinal data is an intrinsic ingredient in developing incisive assessment criteria. If it is accurate, longitudinal data provides the ability to assess the COIN strategy and even more significantly to show the subtle types of progress that lead to success. This is absolutely crucial for gaining and maintaining domestic and international support for the COIN. Intergovernmental organizations, nongovernmental organizations, and individual nations are reluctant to make monetary donations or provide assistance if progress cannot be articulated. Longitudinal assessment tools in a COIN are difficult to set up because they must measure effects over multiple periods of time in a continuously changing operational environment. However, if constructed properly longitudinal assessment is invaluable.

During the Vietnam War the United States experimented with a number of methodologies to determine Vietnamese popular attitudes. However, it was not until January 1967 that the United States began in earnest to institutionalize assessment. This was almost four years after the 2 January 1963 Battle of Ap Bac, which signaled the United States' ever-growing involvement in Vietnam. In 1967 the Defense Department instituted the Hamlet Evaluation System (HES) to measure pacification. The intent of the HES was to construct a device to measure the status of the US-led Revolutionary Development Program in over 12,000 Vietnamese hamlets. In October 1969 the US Pacification Studies Group developed the Pacification Attitudinal Analysis System (PAAS) to depict rural South Vietnamese attitudes toward political and economic development and security. This was followed up in March 1971 with efforts to capture urban Vietnamese perspectives. Both the HES and PAAS had developmental problems that are commonly associated with an evolving assessment system. They struggled with what information to collect and how best to collect it. Additionally, as the HES and PAAS changed their methodologies, what they were measuring changed, which artificially limited the validity of their data. For example, in 1971 the HES changed its scoring system to give greater weight to political factors affecting security.[36] This in turn had an impact on what constituted insurgent forces and how security was measured.

In spite of their limitations, the HES and PAAS experience offer numerous lessons. One of the key lessons is to have an assessment plan when COIN operations begin. When the United States experimented with the HES and PAAS in Vietnam, data collection and automated multivariate regression analysis were in their infancy. Additionally, computer automation was virtually in the "stone age." This factor severely inhibited US assessment efforts and created a delay in developing a comprehensive analysis system. Being unable to holistically assess the insurgency during the early days of US involvement fostered an overreliance on kinetic options that may have worked against the COIN efforts. It also wedded assessment to a dependence on friendly and enemy casualty data to measure success.[37] Another lesson from the HES and PASS era is to develop an understanding of how strategic end-state conditions, objectives, effects, and COGs form the core of longitudinal assessments. If the COG is governing authority, the objectives and effects should assess the progress of the host nation's governing authority and the effectiveness of COIN actions taken to neutralize the insurgent's governing authority. When the HES and PAAS came into being, there was no real longitudinal assessment plan. As a result, when these systems began measuring different variables and defining insurgents differently, the data could not accurately reflect the potential causality of long-term effects.

Perhaps one of the greatest lessons derived from the HES and PAAS experience is that two factors—development and security—accounted for 95 percent of the common variance in the HES data. The development factor indicated 50 to 60 percent of the common variance, and security represented 40 to 50 percent. Analysis of the HES security data also revealed strong correlations between hamlet security and the Viet Cong's ability to tax and recruit.[38]

Today COIN forces have access to a sophisticated array of assessment tools, postmodern technology, and the lessons of the past. However, assessment in COIN operations is still lagging. The most prominent problems encountered in Afghanistan are attitudinal, cultural, and human, and commanders show a surprising passivity to collecting information that is not enemy driven.[39] Overlooked in contemporary COIN operations is the potential validity of the HES's common variance correlations concerning development and security. Numerous International Security Assistance Force organizations are collecting data on both of these factors; however, the data is not providing commanders or political leaders the assessment mechanisms they require to make critical decisions. Host-nation attitudinal data provides invaluable insight into the population's perception of progress and its support for the governing authority. This data may also yield a tilt line indicating the minimum support from the population necessary for success. The concept of the tilt line is to provide the commander a decision tool for modifying or changing the COIN strategy. Figure 59, derived from unscientific ABC News/BBC/ARD poll data on Afghans' positive perceptions of local conditions, presents a theoretical illustration of this concept. The data, similar to the HES, orients on longitudinal analysis of living conditions and security, where most of the common variance for assessing COIN operations traditionally exists. The dotted line is a trend line indicating where the population's support for the Afghan governing authority should be maintained to meet the minimum conditions necessary for success.

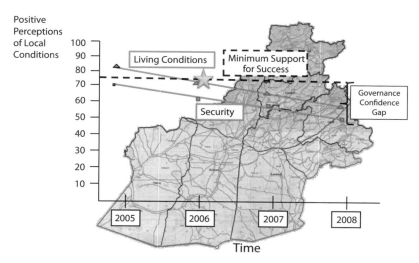

Figure 59. Measuring the governance confidence gap

When perceptions of living conditions and security go above or dip below this line, the perceptions produce a benchmark to measure confidence in the Afghan governing authority. In figure 59, there is a clear indication of a precipitous decline in confidence in the governing authority beginning in 2006. As the confidence gap grows, mounting evidence suggests that the strategy may need to be modified to ensure success. If this data were more scientific and had been analyzed against other environmental factors, it might have shown a clear decision point for the commander.

The intent of this example is not to state that living conditions and security are the sole determining factors in a COIN. Data analysis requires a 360-degree examination of the operational environment to ensure that something as simplistic as a drought or flood has not corrupted the data and implied findings. What this example is trying to show is that commanders need better decision-making tools to support the demands of multifaceted COIN operations.

Conclusion

Design is on the verge of a doctrinal breakthrough that will have a far-reaching impact on how JFCs and their staffs develop campaigns. Forging design's future, however, requires a methodology that can be explained, understood, and applied. This methodology must also incorporate the foresight to visualize the salient differences between the design used for major conventional operations and the design employed in irregular warfare. These radically different operational contexts require distinctive methodological approaches that distill clarity for commanders and staffs. Regardless of the operational context, however, operational design's effectiveness revolves around linking design to decision making and risk analysis. In the future, planning, decision, and execution cycles will be significantly compressed. JFCs and their staffs will face complex networks of nation-state and non-nation-state adversaries, a proliferation of critical technology, crippling cyber attacks, and denial of the electromagnetic spectrum. Overcoming these challenges mandates a vision that will expedite effective decision making and institutionalize approaches to accurately assessing risk. When this occurs, well-developed plans will survive first contact.

Notes

1. Clausewitz, *On War*, 87.
2. Metz, "Counterinsurgent Campaign Planning," 61.
3. Joint Publication (JP) 3-24, *Counterinsurgency Operations*, GL-6.
4. Ibid., GL-5.
5. Jones, *Counterinsurgency in Afghanistan*, 10.
6. Galula, *Counterinsurgency Warfare*, 55.
7. Combined Intelligence Center Vietnam, *The Enemy System*, 2–3.
8. Pike, *Viet-Cong Strategy of Terror*, 17–18.
9. Combined Intelligence Center Vietnam, *The Enemy System*, 2–3.
10. US Department of Defense, *Southeast Asia Statistical Summary*, Table 3, 14 February 1973.
11. Thayer, *War without Fronts*, 32.
12. McFate and Jackson, "Object beyond War," 13.
13. Galula, *Counterinsurgency Warfare*, 54–56.
14. Blackledge, Lardner, and Riechmann, "After Years of Rebuilding, Most Afghans Lack Power."
15. Trofimov, "U.S. Rebuilds Power Plant."
16. Cloud, Rodriguez, and King, "Pakistan Closes Border Crossing."
17. Rising, "Pakistan to Reopen Border Crossing Used by NATO."
18. Kilcullen, *Accidental Guerrilla*, 70–109.
19. US Department of Defense, *Report on Progress toward Security and Stability in Afghanistan and United States Plan for Sustaining the Afghanistan National Security Forces*, 34.
20. Galula, *Counterinsurgency Warfare*, 25.
21. JP 5-0, *Joint Operation Planning*, revision final coordination.
22. Livingston, Messera, and O'Hanlon, *Afghanistan Index*, 13.
23. Gibb, "Where Will All the Soldiers Go?," 47–48.
24. Manwaring and Prisk, *El Salvador at War*, 357.
25. Bacevich et al., *American Military Policy in Small Wars*, 40.
26. Rubin and Farrell, "Awakening Councils by Region."
27. O'Hanlon and Livingston, *Iraq Index*, 3–12.
28. Ibid., 3.
29. *DePetris*, "Remember the Sons of Iraq?"
30. Harari, "Uncertain Future for the Sons of Iraq," 3.
31. Thayer, *War without Fronts*, 157.
32. US Department of Defense, *Southeast Asia Analysis Report*, 32–34.
33. Oriofsky, "Operation Well Being Begins," 448.
34. Craves, "Foreign Assistance Resource Problems," 26.
35. Americas Watch Committee and Lawyers Committee for International Human Rights, *Free Fire*, 4.
36. Thayer, *War without Fronts*, 145–51.
37. Bjelajac, "Guidelines for Measuring Success in Counterinsurgency," 1.
38. Sweetland, *Item Analysis of the HES (Hamlet Evaluation System)*, 3.
39. Flynn, Pottinger, and Batchelor, *Fixing Intel*, 4.

Abbreviations

AA	avenue of approach
ACC	air component commander
ANA	Afghan National Army
ANP	Afghan National Police
ANSF	Afghan National Security Forces
AOR	area of responsibility
APOD	aerial port of debarkation
ARF	Alliance Reaction Force
ATO	air tasking order
AU	Aurigan
AUR	country of Auriga
AUTH	authority
BCT	brigade combat team
Bde	brigade
BLA	Batari Liberation Army
Bn	battalion
BP	branch plan
C2	command and control
C4I	command, control, communications, computers, and intelligence
C4ISR	command; control; communications; computers; and intelligence, surveillance, and reconnaissance
CCF	Chinese communist forces
CCIR	commander's critical information requirement
CHOP	change of operational control
CIVCAS	civilian casualty
CJOA	combined joint operations area
CJTF	combined joint task force
CMH	Center of Military History
COA	course of action
COG	center of gravity
COIN	counterinsurgency
CPVF	Chinese People's Volunteer Force
Def.	defensive
DMZ	demilitarized zone
DP	decision point

DPRK	Democratic People's Republic of Korea
DSM	decision support matrix
DST	decision support template
FDO	flexible deterrent option
FEBA	forward edge of the battle area
FFIR	friendly-force information requirement
FMLN	Farabundo Martí National Liberation Front
FOC	full operational capability
FOM	freedom of movement
GAO	Government Accountability Office
GIROA	Government of the Islamic Republic of Afghanistan
HA	humanitarian assistance
HES	Hamlet Evaluation System
HIG	Hizb-l Islami Gulbuddin
HIK	Hezb-e-Islami Khalis
HQN	Haqqani network
IA	international airport
ICOS	International Council on Security and Development
IJC	International Security Assistance Force Joint Command
INS	insurgent
IO	information operations
ISAF	International Security Assistance Force
ISI	Inter-Services Intelligence (Pakistan)
JFC	joint force commander
JFLCC	joint force land component commander
JIACG	joint interagency coordination group
JIPOE	joint intelligence preparation of the operational environment
JOA	joint operations area
JOPP	joint operation planning process
JP	Joint Publication
JPG	joint planning group
JPME	joint professional military education
KTD	key terrain district

LCC	land component commander
LNO	liaison officer
LOC	line of communications
LOE	line of effort
LOO	line of operation
MA	military action
MACV	Military Assistance Command Vietnam
MCC	maritime component commander
MCO	major combat operation
MEB	Marine expeditionary brigade
MILDEC	military deception
NAI	named area of interest
NEO	noncombatant evacuation operation
NGO	nongovernmental organization
NLF	National Liberation Front
NVA	North Vietnamese Army
OA	objective area
OEF	Operation Enduring Freedom, operational exploitation force
Off.	offensive
OGA	other government agency
OIF	Operation Iraqi Freedom
OPFOR	opposing force
OPLAN	operation plan
Ops	operations
PAAS	Pacification Attitudinal Analysis System
PDE	planning-decision-execution
PER	country of Perseus
PF	popular forces
PIR	priority intelligence requirement
PMESII	political, military, economic, social, information, and infrastructure
POTF	psychological operations task force
QTB	Quetta Shura Taliban
RC	regional command
RCT	regimental combat team
RF	regional forces

RIP/TOA	relief in place / transfer of authority
ROL	rule of law
RSOI	reception, staging, onward movement, and integration
SA	situational awareness
SAG	country of Sagitta
SOCC	special operations component commander
SOF	special operations forces
SOI	Sons of Iraq
SP	sequal plan
SPOD	seaport of debarkation
SPOE	seaport of embarkation
TAA	tactical assembly area
TAI	target area of interest
TBM	theater ballistic missile
TOA	transfer of authority
TRADOC	Training and Doctrine Command
TTP	Tehrik-i-Taliban
UN	United Nations
UNC	UN Command
UNSCR	UN Security Council resolution
USAID	United States Agency for International Development
USCENTCOM	US Central Command
VC	Viet Cong
VCI	Viet Cong Infrastructure
VP	country of Vulpecula
Vul	Vulpeculan
WARNORD	warning order
WMD	weapons of mass destruction

Bibliography

Afrasiabi, Kaveh, and Abbas Maleki. "Iran's Foreign Policy after 11 September." *Brown Journal of World Affairs* 9, no. 2 (Winter–Spring 2003): 255–65.

Americas Watch Committee and Lawyers Committee for International Human Rights. *Free Fire: A Report on Human Rights in El Salvador*, Fifth Supplement. New York: Americas Watch Committee and Lawyers Committee for International Human Rights, August 1984.

Bacevich, A. J., James D. Hallums, Richard H. White, and Thomas F. Young. *American Military Policy in Small Wars: The Case of El Salvador.* Washington, DC: Pergamon-Brassey's, 1988.

Badeau, Adam. *Military History of Ulysses S. Grant from April 1861 to April 1865.* Vol. 1. New York: D. Appleton, 1885.

Bartlett, F. C. *Remembering: An Experimental and Social Study.* Cambridge: Cambridge University Press, 1932.

Bearss, Edwin Cole. *The Vicksburg Campaign.* Vol. 1, *Vicksburg Is the Key.* Dayton, OH: Morningside House, 1985.

Belasco, Amy. *The Cost of Iraq, Afghanistan, and Other Global War on Terror Operations since 9/11.* Congressional Research Service, 29 March 2011.

Bjelajac, S. N. "Guidelines for Measuring Success in Counterinsurgency." Paper presented at the Special Warfare and Incipient Insurgency Working Groups at the XVIII MORS [Military Operations Research Society]. John F. Kennedy Special Warfare Center, Fort Bragg, NC, 19–21 October 1966.

Blackledge, Brett J., Richard Lardner, and Deb Riechmann. "After Years of Rebuilding, Most Afghans Lack Power." *Associated Press*, 19 July 2010.

Bush, Pres. George W. *President's Address to a Joint Session of Congress and the American People.* Washington, DC: White House, 20 September 2001.

Catton, Bruce. *Grant Moves South.* Boston: Little, Brown, 1960.

Chapius, Oscar. *A History of Vietnam from Hong Bang to Tu Duc.* Westport, CT: Greenwood Press, 1995.

Ch'oe, Yong-ho, Peter H. Lee, and William Theodore de Bary, eds. *Sources of Korean Tradition II: From the Sixteenth to the Twentieth Centuries.* New York: Columbia University Press, 2000.

Clausewitz, Carl von. *On War*. Edited and translated by Michael Howard and Peter Paret. Princeton, NJ: Princeton University Press, 1989.

Cloud, David S., Alex Rodriguez, and Laura King. "Pakistan Closes Border Crossing, Says NATO Copters Killed 3 of Its Soldiers." *Los Angeles Times*, 30 September 2010.

Combined Intelligence Center Vietnam. *The Enemy System: Binh Thuan Province*. Research and Analysis Study ST 71-02, 6 July 1971. In *Records of the Military Assistance Command Vietnam*. Part 2, *Classified Studies from the Combined Intelligence Center Vietnam, 1965–1973*, edited and compiled by Robert E. Lester. Microfilm from the holdings of the Library of the US Army Military History Institute, Carlisle Barracks, PA.

———. *The Enemy System: Subregion 1*. Research and Analysis Study ST 70-01, 24 August 1970. In *Records of the Military Assistance Command Vietnam*. Part 2, *Classified Studies from the Combined Intelligence Center Vietnam, 1965–1973*, edited and compiled by Robert E. Lester. Microfilm from the holdings of the Library of the US Army Military History Institute, Carlisle Barracks, PA.

Craves, John P. "Foreign Assistance Resource Problems and Requirements for Low Intensity Conflict." Monograph. Fort Leavenworth, KS: School of Advanced Military Studies, May 1986.

D'Amura, Ronald M. "Campaigns: The Essence of Operational Warfare." *Parameters* (US Army War College) 17, no. 2 (Summer 1987): 42–51.

de Czege, Huba Wass. "Systemic Operational Design: Learning and Adapting in Complex Missions." *Military Review*, January–February 2009, 2–13.

de Hartog, Leo. *Genghis Khan: Conqueror of the World*. New York: Barnes & Noble, 1999.

DePetris, Daniel R. "Remember the Sons of Iraq? A Review of Michal Harari's SoI Briefing." *Small Wars Journal* (blog), 7 November 2010. http://smallwarsjournal.com/blog/2010/11/remember-the -sons-of-iraq/.

Dickson, Keith D. "Operational Design: A Methodology for Planners." *Journal of the Department of Operational Art and Campaigning*, Joint Advanced Warfighting School, Spring 2007, 23–38.

Flynn, Maj Gen Michael T., Capt Matt Pottinger, and Paul D. Batchelor. *Fixing Intel: A Blueprint for Making Intelligence Relevant in Af-*

ghanistan. Washington, DC: Center for a New American Security, 5 January 2010.

Fuller, J. F. C. *A Military History of the Western World*. Vol. 1. 1954. Reprint, New York: Da Capo Press, 1987.

Gabriel, Richard A., and Donald W. Boose, Jr. *The Great Battles of Antiquity: A Strategic and Tactical Guide to Great Battles That Shaped the Development of War*. Westport, CT: Greenwood Press, 1994.

Galula, David. *Counterinsurgency Warfare: Theory and Practice*. Westport, CT: Praeger Security International, 1964.

Gibb, Thomas. "Where Will All the Soldiers Go?" *US News and World Report*, 22 October 1990, 47–48.

Gordon, Michael R., and Bernard E. Trainor. *The Generals' War*. New York: Little, Brown and Co., 1995.

Grant, R. G., ed. *1001 Battles That Changed the Course of World History*. New York: Universe Publishing, 2011.

Greene, Robert. *The 33 Strategies of War*. New York: Penguin, 2007.

Harari, Michal. "Uncertain Future for the Sons of Iraq." Background paper for the Institute for the Study of War, 3 August 2010. http://www.understandingwar.org/files/Backgrounder_SonsofIraq.pdf.

International Council on Security and Development (ICOS). *Afghanistan Transition: Missing Variables*. London: ICOS, November 2010.

Jian, Chen. "The Sino-Soviet Alliance and China's Entry into the Korean War." Working paper no. 1. Washington, DC: Woodrow Wilson Center's Cold War History Project, 1991.

Johnson, Robert U., and Clarence C. Buel, eds. *Battles and Leaders of the Civil War*. Vol. 2, *The Struggle Intensifies*. New York: Century Company, 1888.

Joint Publication 1-02. *Department of Defense Dictionary of Military and Associated Terms*, 8 November 2010 (as amended through 31 December 2010).

Joint Publication 3-0. *Doctrine for Joint Operations*, 1 February 1995.

Joint Publication 3-0. *Joint Operations*, 17 September 2006.

Joint Publication 3-0. *Joint Operations*, 11 August 2011.

Joint Publication 3-24. *Counterinsurgency Operations*, 5 October 2009.

Joint Publication 5-0. *Joint Operation Planning*, 26 December 2006.

Joint Publication 5-0. *Joint Operation Planning*, revision final coordination, 25 October 2010.

Joint Publication 5-0. *Joint Operation Planning*, 11 August 2011.

Joint Warfighting Center Doctrine Pamphlet 10. *Design in Military Operations: A Primer for Joint Warfighters*, 20 September 2010.

Jones, Seth G. *Counterinsurgency in Afghanistan.* RAND Counterinsurgency Study. Vol. 4. Santa Monica, CA: 2008.

Kapsis, James E. "The Failure of U.S.-Turkish Pre-War Negotiations: An Overconfident United States, Political Mismanagement, and a Conflicted Military." *Middle East Review of International Affairs* 10, no. 2 (September 2006). http://meria.idc.ac.il/journal/2006/issue3/jv10no3a3.html.

Kem, Jack D. *Campaign Planning: Tools of the Trade.* 2nd ed. Fort Leavenworth, KS: Department of Joint and Multinational Operations, US Army Command and General Staff College, June 2006.

Kilcullen, David. *The Accidental Guerrilla: Fighting Small Wars in the Midst of a Big One.* Oxford: Oxford University Press, 2009.

Li, Xiaobing, Allan R. Millet, and Bin Yu, trans. and ed. *Mao's Generals Remember Korea.* Lawrence, KS: University Press of Kansas, 2000.

Livingston, Ian S., Heather L. Messera, and Michael O'Hanlon. *Afghanistan Index: Tracking Variables of Reconstruction & Security in Post-9/11 Afghanistan.* Washington, DC: Brookings, 31 July 2010.

Livingston, Ian S., and Michael O'Hanlon. *Iraq Index: Tracking Variables of Reconstruction & Security in Post-Saddam Iraq.* Washington, DC: Brookings, 31 October 2010.

Maleki, Abbas. "Decision Making in Iran's Foreign Policy: A Heuristic Approach." http://www.caspianstudies.com/article/Decision%20Making%20in%20Iran-FinalDraft.pdf.

Manwaring, Max G., and Court Prisk, eds. *El Salvador at War: An Oral History of Conflict from the 1979 Insurrection to the Present.* Washington, DC: National Defense University Press, 1988.

McFate, Montgomery, and Andrea V. Jackson. "The Object beyond War: Counterinsurgency and the Four Tools of Political Competition." *Military Review*, January–February 2006, 13–26.

McPherson, James. *Battle Cry of Freedom.* New York: Oxford University Press, 1988.

Metz, Steven. "Counterinsurgent Campaign Planning." *Parameters* 19, no. 3 (September 1989): 60–68.

Mintz, Alex. "How Do Leaders Make Decisions?: A Polyheuristic Perspective." *Journal of Conflict Resolution* 48, no. 1 (1 February 2004): 3–13.

Mora, Edwin. "Half of Afghan Military Forces Won't Achieve 1st Grade Literacy Level by 2012." *CNSNews.com*, 9 May 2011. http://www.cnsnews.com/news/article/half-afghan-forces-expected-master-1st-g.

National Security Directive 54. *Responding to Iraqi Aggression in the Gulf* (U), 15 January 1991.

Nelson, Harold H. "The Battle of Megiddo." PhD diss., University of Chicago, 1913.

O'Hanlon, Michael E., and Ian Livingston. *Iraq Index: Tracking Variables of Reconstruction & Security in Post-Saddam Iraq*. Washington, DC: Brookings, 31 October 2010.

Oriofsky, Stephen, ed. "Operation Well Being Begins." *Facts on File* 43, no. 17 (June 1983): 448.

Piaget, Jean, and Bärbel Inhelder. *Memory and Intelligence*. London: Routledge and Kegan Paul, 1973.

Pike, Douglas. *The Viet-Cong Strategy of Terror*. Prepared for the United States Mission, Viet-Nam, February 1970. In Indochina Archive, Washington, DC, edited by Douglas Pike. Accessed 1 March 2010, http://www.vietnam.ttu.edu/star/images/231/23104 02003a.pdf.

———. *War, Peace, and the Viet Cong*. Cambridge, MA: MIT Press, 1969.

Pogue, Forrest C. *The Supreme Command*. Center of Military History (CMH) Publication 7-1. Washington, DC: CMH, US Army, 1996.

Pritchard, James B. *Ancient Near Eastern Texts Relating to the Old Testament*. 3rd ed. Princeton, NJ: Princeton University Press, 1969.

Qiao, Liang, and Wang Xiangsui. *Unrestricted Warfare*. Beijing: PLA Literature and Arts Publishing House, February 1999.

Rising, David. "Pakistan to Reopen Border Crossing Used by NATO." *Washington Times*, 9 October 2010.

Rubin, Alissa J., and Stephen Farrell. "Awakening Councils by Region." *New York Times*, 22 December 2007.

Sears, Stephen W. *George B. McClellan: The Young Napoleon*. New York: Da Capo Press, 1988.

Stackpole, Edward J. *From Cedar Mountain to Antietam*. Mechanicsburg, PA: Stackpole Books, 1959.

Strange, Joseph. *Centers of Gravity and Critical Vulnerabilities: Building on the Clausewitzian Foundation So We Can All Speak the Same Language*. Perspectives on Warfighting Series, no. 4, 2nd ed. Quantico, VA: Marine Corps Association, 1996.

Sweetland, Anders. *Item Analysis of the HES (Hamlet Evaluation System)*. Santa Monica, CA: RAND, 1968.

Thayer, Thomas C. *War without Fronts: The American Experience in Vietnam*. Boulder, CO: Westview Press, 1985.

Training and Doctrine Command (TRADOC) Pamphlet 525-5-500. *Commander's Appreciation and Campaign Design*, version 1.0, 28 January 2008.

Trofimov, Yaroslav. "U.S. Rebuilds Power Plant, Taliban Reap a Windfall." *Wall Street Journal*, 13 July 2010.

US Army. *Army Strategic Planning Guidance '99* (draft). 15 February 1999.

US Army. *On Point: U.S. Army in Operation Iraqi Freedom*. Washington, DC: Office of the Chief of Staff, 2004.

US Army War College, Department of Military Strategy, Planning, and Operations. *Campaign Planning Handbook, Final Working Draft AY 08*. Carlisle, PA: US Army War College, 2008.

———. *Campaign Planning Primer AY 07*. Carlisle, PA: US Army War College, 2007.

US Department of Defense (DOD). *2008 National Defense Strategy*. Washington, DC: DOD, June 2008.

US Department of Defense. *Report on Progress toward Security and Stability in Afghanistan* (report to Congress in accordance with section 1230 of the National Defense Authorization Act for Fiscal Year 2008 [Public Law 110-181], as amended) and *United States Plan for Sustaining the Afghanistan National Security Forces* (report to Congress in accordance with section 1231 of the National Defense Authorization Act for Fiscal Year 2008 [Public Law 110-181]). Submitted to Congress 28 April 2010. Revised version submitted 21 May 2010.

US Department of Defense. *Report on Progress toward Security and Stability in Afghanistan and United States Plan for Sustaining the Afghanistan National Security Forces*. April 2011.

US Department of Defense, Office of the Assistant Secretary of Defense, Comptroller. *Southeast Asia Statistical Summary*, Table 3, 14 February 1973.

US Department of Defense, Office of the Assistant Secretary of Defense, Systems Analysis. *Southeast Asia Analysis Report*, August–October 1971.

US Department of Defense press briefing. Secretary of Defense Donald H. Rumsfeld and General Richard Myers. Subject: Operation Enduring Freedom, 7 October 2001.

US Government Accountability Office (GAO). *Afghanistan Security: Afghan Army Growing, but Additional Trainers Needed; Long-term Costs Not Determined*. Report to Congressional Addressees. Washington, DC: GAO, January 2011.

US Joint Forces Command, Joint Center for Operational Analysis. *Operation Iraqi Freedom May 2003 to June 2004: Stabilization, Security, Transition, and Reconstruction in a Counterinsurgency (Part Two)*. Norfolk, VA: Joint Forces Command, January 2006.

Wells, Peter S. *The Battle That Stopped Rome: Emperor Augustus, Arminius, and the Slaughter of the Legions in the Teutoburg Forest*. New York: W. W. Norton, 2003.

Wertheimer, Max. "Gestalt Theory." An address before the Kant Society, Berlin, 7 December 1924, Erlangen, Germany. In *Source Book of Gestalt Psychology*, trans. Willis D. Ellis. New York: Harcourt, Brace and Co, 1938.

Wolters, O. W. "On Telling a Story of Vietnam in the Thirteenth and Fourteenth Centuries." *Journal of Southeast Asian Studies* 26, no. 1 (1995): 63–74.

Index

National Liberation Front, 68, 105. *See also* North Vietnamese Army and Viet Cong
National Military Strategy, 33
National Security Strategy, 33
NATO, 8, 93
North Korea, 7, 11, 54, 56. *See also* Democratic People's Republic of Korea
North Vietnamese Army, 68, 86. *See also* National Liberation Front and Viet Cong

Obama, Pres. Barack 13
objectives, 6, 9, 13, 19, 21, 25, 27, 31–42, 47–53, 60–64, 68, 74, 78, 83, 86, 88–89, 93, 96–97, 102–3, 105, 111, 114
observed system, 6–7, 9
Operation Desert Shield, 34, 63
Operation Desert Storm, 34, 44–47, 63
Operation Enduring Freedom (OEF), 10, 13, 34, 73–74
Operation Iraqi Freedom (OIF), 7, 11, 57
Operation Overlord, 49
operation plan (OPLAN), 24, 55, 57
operational art, 21, 23–28
operational environment, 5–6, 15, 18, 36–37, 41–43, 50, 64, 69, 71, 76, 88–91, 93, 105, 113, 116
operational planning, 24
Organización Democrática Nacionalista (ORDEN), 110

Pacification Attitudinal Analysis System (PAAS), 113–14
Pakistan, 93
Peninsula Campaign (US Civil War), 2–3, 5
Persian Gulf, 34
phases: of an operation, 21, 31, 34, 49–53, 55, 64–65, 74, 76, 83, 93, 95–97; of a war game, 65, 69–70
Pike, Douglas, 85–86
planning-decision-execution cycle, 110
PMESII, 6, 86, 88

Qiao, Liang, 9–10
Quetta Shura Taliban, 93

Republic of Korea, 7, 11
Revolutionary Development Program, 113
Richmond, VA (in US Civil War), 2–5
RIP/TOA, 96, 100, 102
risk analysis, 5, 12, 116
Rumsfeld, Donald, 34

Saudi Arabia, 34
Second Battle of Manassas (US Civil War), 5
September 11, 2001. *See* 9/11
Seven Days' Battles (US Civil War), 4
Shah Ala al-Din Muhammad II, 62–63
Shiites, 93, 106
South Korea. *See* Republic of Korea
South Vietnam, 69, 86, 108, 113. *See also* Vietnam
Southwest Pacific, 47–48
Strange, Joe, 42
structure (in operational design), 1, 25, 27, 59, 65, 69, 76, 97, 100
Sunnis, 93, 107
sustainment, 63
synchronization matrix, 71–72, 111
systems analysis, 40–41, 85–88

Taliban, 8, 13, 34, 89, 93
target area of interest (TAI), 71–75
tasks (and operational design), 32, 37–39, 44, 47–48, 52–53, 73–74, 102–3
termination criteria, 11, 33, 35–37, 62
Tet offensive, 68
Teutoburger, 54
Thayer, Thomas, 109
Thutmose III, 21–23
time (as a factor in operational design), 9–12, 51, 56, 65, 69, 84, 96
Torkham Gate, 93
Transoxiana, 62–63
Turkey, 11–12

Union forces (US Civil War), 2–5, 78–79
United Nations (UN), 52–53, 56
United Nations Security Council, 8, 33, 35, 37
United States Agency for International Development (USAID), 89
unrestricted warfare, 9–10